Anthropology students increasingly need a quantitative background, but statistics are often seen as difficult and impenetrable. *Statistics for Anthropology* offers students of anthropology and other social sciences an easy, step-by-step route through the statistical maze. In clear, simple language, using relevant examples and practice problems, it provides a solid footing in basic statistical techniques, and is designed to give students a thorough grounding in methodology, and also insight into how and when to apply the various processes.

The book assumes a minimal background in mathematics, and is suitable for the computer literate and illiterate. Although only a hand calculator is needed, computer statistical software can be used to accompany the text. The book will be a 'must-have' for all anthropology and social science students needing an introduction to basic statistics.

Statistics for Anthropology

Statistics for Anthropology

LORENA MADRIGAL

Associate Professor, Department of Anthropology
University of South Florida, Tampa, FL 33620, USA

PUBLISHED BY THE PRESS SYNDICATE OF THE UNIVERSITY OF CAMBRIDGE
The Pitt Building, Trumpington Street, Cambridge CB2 1RP, United Kingdom

CAMBRIDGE UNIVERSITY PRESS
The Edinburgh Building, Cambridge CB2 2RU, United Kingdom
40 West 20th Street, New York, NY 10011-4211, USA
10 Stamford Road, Oakleigh, Melbourne, 3166, Australia

First published 1998

Printed in the United Kingdom at the University Press, Cambridge

Typeset in Utopia 9.25/13.5 and MetaPlus, in Quark XPress™   [SE]

A catalogue record for this book is available from the British Library

ISBN 0 521 57116 2 hardback
ISBN 0 521 57786 1 paperback

To Steve and Sofía

CONTENTS

This book was written with the desire to fill a void in statistics textbooks. It is written for the social science student in general, but the anthropology student in particular. My experience and that of other anthropologists who teach statistics has been that most social science textbooks do not consider the tremendous diversity of interests which concerns anthropologists. The type of examples and exercises used in a textbook are very important, if a student is to be challenged by and interested in learning quantitative methods. While not written to the exclusion of other disciplines in the social sciences, the primary focus of this book is that of anthropology, including as many of its different facets as was possible.

This book is published at a time when the need for a solid quantitative education is becoming more important for anthropologists in all fields of study. Indeed, sophisticated quantitative tests are becoming extremely frequent in the biological anthropology and archaeological journals. The growing importance of statistical education in anthropology was made apparent by a survey of physical anthropologists performed during 1990–1991 (Wienker and Bennett, 1992). The survey asked members of the American Association of Physical Anthropologists to list their most important academic deficiencies. Statistics was ranked as the number 1 deficiency in their training.

The purpose of this book is to educate students on basic univariate tests, not techniques such as multivariate analysis or multiple regression. Once univariate statistics are mastered, students will be in a position to take higher-level statistical classes inside or outside of their own discipline. This book intends to provide a solid basic footing from which the student can learn more, and can critically read research papers. Therefore, it is intended for either high-level undergraduates, or first-year graduate students taking their first statistics course.

The book assumes the student has a small (if not minimal) background in mathematics. Thus, especially during the first chapters, all mathematical computations are shown step by step. This book considers both the computer-literate and non computer-literate student. It can be read and used with nothing more sophisticated than a hand-calculator, or it can be accompanied by a computer package. To illustrate the great advantage and ease of computer use, the program SAS/ASSIST was chosen. Appendix A provides a brief review of

SAS/ASSIST, which should be read if it is to be used in conjunction with the text. Students who do not use a computer, or who use another computer package, will find the SAS/ASSIST examples of use, however.

Most chapters have the same format: usually every statistical test will first be discussed, then illustrated with an example. A practice problem follows; the reader may skip it if desired. A boxed example of the SAS/ASSIST output is next, followed by a research example of the use of the particular test. The list of key concepts summarizes the main points covered in the chapter, providing topics for study. Exercises are also included, the answers to some of which appear at the end of the book.

In order to remain gender neutral, I chose to alternate the use of masculine and feminine pronouns through the book.

The publishers who granted copyright permission to reproduce information in the statistical tables are acknowledged at the foot of the tables. Here, I wish to acknowledge the *Journal of Heredity* for permitting me to reproduce a picture (Figure 4.2) which appeared in 'Corn and men', by A. F. Blakeslee (1914), volume 5 (11): 512. I also wish to acknowledge the Regents of the University of California, for allowing me to reproduce a table from *Banana Fallout* by Trevor Purcell (1993), published by the UCLA Center for African American Studies. The Statistical Analysis System (SAS) Institute is acknowledged for its support and permission to use the SAS package in the book.

I wish to acknowledge Dr Tracey Sanderson, Commissioning Editor, Biological Sciences, Cambridge University Press for her help during the various stages of the production of this book. I would also like to acknowledge the faculty of the Department of Anthropology, University of South Florida, for their encouragement, and Dr N. White in particular for offering some of her data. Revisions to the manuscript by Ms Trina Thompsom and Mr Steve Allison are gratefully acknowledged, although the author takes full responsibility for the book. Mrs Maggie diPietra provided excellent graphical advise and artistry in the form of several figures. The help provided by the staff of the interlibrary loan service of the University of South Florida is also acknowledged.

1 Introduction to statistics

In this chapter the reader is introduced to the field of statistics. The chapter has three main sections: the relationship between statistics and science, basic definitions, and statistical notation and rules of rounding. Throughout the text **bold** type is used for the first occurrence of specialised/statistical terms (and for comments annotating parts of computer output), and *italic* is used for emphasis.

1.1 Statistics and scientific inquiry

Statistics are usually defined as facts and figures which summarize a data set, or the methods utilized to achieve such figures. But, as R. A. Fisher explains in his seminal book *Statistical Methods for Research Workers* (1991a), statistics are also the study of populations (to be defined below) and their variation. It is precisely because we wish to understand the variation present in the population of our interest that we use figures (statistics) to describe the sample with which we work.

Statistics are an integral part of science. Not only do they help the researcher to organize and interpret data, but they are also involved in the process of hypothesis testing, perhaps the main difference between science and other forms of knowledge pursuits.

Although the concepts of theory and hypothesis are extremely important in science, they are frequently defined differently in various disciplines. In this book, we define a **theory** as a set of hypotheses that has been tested repeatedly and not rejected. A **hypothesis** is defined as an explanation of observed facts, an explanation which may or may not be testable. For a hypothesis to be scientific, it must be presented in a testable form. In other words, scientific hypotheses should be stated so that they are capable of being rejected by empirical evidence.

For example, the theory of evolution includes many hypotheses about the nature of population genetic change. A well-documented evolutionary change is that of the color change of peppered moths in London during the height of the industrial revolution. Before the onset of heavy pollution in London, the majority of the moth population was lightly pigmented because light-colored moths were well camouflaged on light tree barks from their predators (birds). However, because of heavy pollution, tree bark became progressively darker. As a result, dark moths became better camouflaged, and the frequency of light moths decreased and dark ones increased. According to Volpe (1985) the frequency of dark peppered moths increased from less than 1% in 1848 to 95% in 1898. With the control of pollution in the 1900s, tree barks became lighter, light moths had the survival advantage, and the frequency of dark moths decreased. These are the observed facts. Various hypotheses can be proposed to explain them. For example, it could be proposed that such changes in moth coloration are the result of a supernatural Being testing our faith. Or it could be hypothesized that birds acted as a natural selection agent, and that the change in moth coloration was an evolutionary change experienced by the moth population. Both propositions are explanations of the facts, but only the second can be empirically tested and therefore be considered scientific.

Please note that a hypothesis can not be proven true. It can however, be rejected. A hypothesis that has not been rejected after many studies has more truth value than one that has been supported by only a single study or none at all. In the case of the peppered moth, the scientific hypothesis presented above has been tested repeatedly, and has not been rejected. Thus, in the scientific world, it is accepted as the most likely explanation for the observed changes in the moth population. Notice that I did not say that it has been proved to be correct. Now, how can statistics be of use in this example? Researchers need an objective, widely recognized and accepted method to decide if the change of frequencies of colors is a *significant* one, or if it could have occurred by pure chance. Statistics allow investigators to decide if this (or any other experimental change) is more likely due to a treatment (in this case, the birds), or more likely due to pure chance. As a tool to test hypotheses, statistics are an integral part of scientific endeavor.

What happens if a scientific hypothesis is rejected? In this case, the theory encompassing the hypothesis needs to be revisited. And this is also a fundamental distinction between science and other forms of human knowledge. Science changes; it is flexible. For example, the theory of evolution has itself evolved through the years. Thus, Ernst Heckel proposed (and Charles Darwin endorsed) that the stages of embryonic development were a replay of the evolutionary stages through which species had evolved ('Ontogeny recapitulates phylogeny'). But this hypothesis has now been rejected. Does that mean that the theory of evolution should be rejected? Not at all! It only means that the theory needs to be revised to accommodate a new hypothesis that explains the facts of embryological development. (For an excellent, accessible account of the embryology of the mammalian ear bones, see Gould, 1990.) As Futuyma (1995: 163) so ably puts it: '...good scientists *never* say they have found absolute "truth."' (original emphasis).

Another link between statistics and science is that science is based on the repeatability of results. Thus, after an initial test, other investigators should be able to repeat, and either refute or replicate the first investigator's conclusions. For example, a few years ago, some investigators claimed that they had achieved cold fusion in the laboratory. Other investigators replicated the experimental procedures, and failed to achieve the same results. Today, the consensus is that cold fusion was not achieved in the initial experiment (Close, 1991). Therefore, it is of fundamental scientific importance that a hypothesis be able to be tested, and that the test be able to be replicated by others. However, anthropological data are not usually collected as a result of an experimental test which can be replicated. In fact, anthropologists can not provide for true repeatability of observations in most cases (how can we replicate a historical event, a culture?). Precisely because of the nature of anthropological studies, researchers must clearly quantify their observations. Imagine that a researcher is interested in studying the presence of property ownership in married females in hunting and gathering groups. After she collects her data in one population, she turns to the literature to compare her results with those of other investigators. Imagine the investigator's frustration if an author reports that '...some married females own property...', whereas another reports that '...a few married females own property...', whereas another writes that '...many married females own property...' Just what exactly is 'some' or 'a few' or 'many'? Is it 40%, 60% ? Because social events can not be replicated in the laboratory, it is of fundamental importance that researchers quantify their results for purposes of comparison with other investigators' results.

Interestingly, Fisher (1991a) notes that the need to repeat experiments (or compare one's findings with others), attests to the fact that scientists are interested in the entire population, not the individual observation. Indeed, if a researcher were to collect data on the relation between marital status and property ownership in females, her research would only truly contribute to the advancement of knowledge when compared with other investigations. Only then would the researcher know if her research supports or contradicts what others have reported.

1.2 Basic definitions

1.2.1 Validity and reliability

When data are collected for a research project, a principal concern should be that the data be valid. Valid data are accurate. **Validity** is the degree to which the method for collecting information results in accurate information. Physical anthropologists who collect anthropometric data (body measurements) try to maximize the degree of validity of their measurements by using the most precise instruments. Thus, whereas an instrument may be able to measure a subject's height as 156cm, another instrument may be able to measure the same subject's height as 156.23cm. The latter measure is more valid (accurate) than is the former. In the case of anthropometry then, a better instrument will result in a more accurate datum. In some other forms of anthropological investigation, however, it may not be as easy to make the measuring instrument more valid. For example, an anthropologist may be interested in studying the reproductive history of post-menopausal females. In that case, he would wish to record (among other things) the females' age at first menstruation (menarche), the age at each pregnancy, the number of pregnancies, and the outcome of each pregnancy (abortion, stillbirth, live birth). If the researcher is working with a population that lacks adequate medical records, he must rely on the female's own account of these events. But how can he be sure that the female did not forget a pregnancy that resulted in a very early abortion, or that she had her menarche at 12, not 13 years? In this and many other examples of anthropological data analysis, the researcher may not be able to show that the data are valid, but at least he may show that the data are reliable.

Reliability is the degree to which observations of a study are repeatable. A measuring instrument is reliable if it generates consistent observations at two points in time. In the case of reproductive history analysis, the researcher could interview the females again, and compare the results of both interviews (Madrigal, 1991). An observation cannot be valid if it is unreliable. Thus, if a female reports a different number of pregnancies in the two interviews, the investigator knows that the data are neither reliable nor valid. If the subject reports the same number of pregnancies in both interviews, the data are reliable, and they are likely to be valid, but the interviewer can not be sure about the latter. It is perfectly possible that the female forgot (or perhaps she never knew) she had that pregnancy that ended in an early abortion, and that she did not report it in either interview. Therefore, an observation can be reliable (consistent) without being valid (a true measure).

Clearly, scientists wish to collect valid information. It is imperative that a measuring instrument (a measuring tape or an open-ended interview) be as accurate as possible, and that data be reliable. (For an excellent source on reliability and validity, see Fleiss, 1986.)

1.2.2 Variables and constants

A **variable** is a characteristic of subjects that differs in value. For example, in a study on income among heads of households in a neighborhood, income is a variable, since it varies (although more than one person may have the same income). The specific **variate** or **datum** is the observation recorded in a subject.

A **constant** is an observation which has the same value across subjects. Thus, if all the household heads interviewed are members of the same family group, then family membership is a constant (it does not vary).

1.2.3 Independent and dependent variables

The **independent variable** is the variable that is manipulated by, or is under the control of, the researcher. This is easy to understand if the research project is taking place in a laboratory. For example, if the researcher is interested in studying the growth of tomato plants, she can manipulate the amount of water given to the plants, and observe the effect of watering on growth. Watering in this example is the independent variable. Needless to say, in a science such as anthropology, such manipulation is difficult to achieve. However, according to the definition above, the independent variable is under the control of the researcher. Let us say that a researcher is interested in the political ideology of members of different socio-economic classes. He can collect data on various social and economic variables and establish his subjects' social class. Then the researcher can collect data on their political ideology. Finally, he can perform a statistical test to determine whether socio-economic status significantly influences political ideology. The key in this example is that the investigator can establish the subjects' socio-economic status, and divide them accordingly. Then he can record (observe) the subjects' political ideology. In this sense, what the researcher is doing in the community is not that different from what the experimenter is doing with tomato plants in the laboratory. The field worker can clearly specify the socio-economic level of the subjects, just as the laboratory investigator can clearly specify how much water goes into the tomato plants. Then the anthropologist can record the subjects' political ideology, just as the laboratory investigator can record the plants' growth.

Clearly, both in the laboratory and community research project, amount of water and socio-economic status are not the variables under study. Plant growth and political ideology are. The **dependent variable** is the variable that is under study, or the variable that the researcher in interested in. The researcher does not control it, but observes it. Thus, the investigator will not sit and argue with his subjects in an attempt to manipulate their political ideology. By the same token, once the laboratory researcher has administered water to the plants, all she can do is record the plants' growth. In general, scientists are interested in studying the impact of the independent variable on the dependent one. For example, the effect of education on women's fertility, of nutritional intake on children's growth, of smoking on fetal development, etc.

It should be emphasized that no matter how 'unscientific' a research project may seem, it can still be designed with clearly stated independent and dependent variables. If so done, the project will allow hypothesis tests about the impact of the independent variable on the dependent one.

1.2.4 Control and experimental groups

As the independent variable is under the control of the researcher, she can divide her subjects into groups according to their independent variable values. In fact, she can manipulate the value of the independent variable so that some subjects receive an experimental 'treatment' and some do not. The **experimental group** receives the treatment, whereas the **control group** does not. Thus, the control group is used for the purpose of making comparisons.

If subjects are to be divided into experimental and control groups, the statistical decision derived from the experiment rests on the assumption that the assignment to groups was done randomly. That is, the researcher must be assured that no uncontrolled factors are influencing the results of the statistical test. The probability laws on which the statistical tests rest assume such randomization. If subjects are randomly assigned to treatment or control groups and the treatment does not have an effect, then the results of the experiment will be determined entirely by chance (Fisher, 1991b) and not by the treatment.

In a non-experimental setting, the researcher must do whatever is possible to achieve randomization, that is, to control bias. If subjects from both groups differ in another variable besides the treatment, then they should be matched so that results are not confounded by this other variable. The purpose of correcting the bias is to minimize the effect of covariates, or compounding variables, on the result of the experiment. In a field study, they may necessitate a large number of controls to be initially selected (Maxwell and Delaney, 1990; Rosenbaum, 1995). The importance of randomization was championed by Fisher (1991b).

1.2.5 Scales of measurement and variables

Anthropologists encounter many different kinds of variables, from ideology to income, from gene frequencies to culture-specific kinship terms. As different statistical techniques are designed for different types of variables, it is important to become familiar with the different forms of variables.

Qualitative variables classify subjects according to the kind or quality of their attributes. Examples of such variables are political party membership, ethnic membership, or even blood type. If an investigator works with qualitative variables, he may code the different variates with numbers. For example, he could assign a number 1 to Thais, a number 2 to Vietnamese, etc. However, just because the data have now been coded with numbers, this does not mean they can be analyzed with just any statistical method. For example, it is possible to report the most common (modal) ethnic membership in a sample, but it is not possible to compute the mean ethnic membership.

Another important point about research with qualitative variables concerns the coding system, which should consist of mutually exclusive and exhaustive categories. Thus, each observation should be placed in one and only one category (mutual exclusiveness), and all observations should be categorized (exhaustiveness).

Ranked or **ordered variables** are variables that clearly have a hierarchy although the distance between the different variates may or may not be set. For example, an anthropologist could interview an informant about the prestige status of various household heads in the community. After the interview, the investigator could proceed to rank-order such individuals. However, when ranking the individuals, the researcher does not imply that the difference in prestige among the various household heads is the same. It is conceivable that the first two individuals in the ranking are very close to each other, whereas the third one is quite distant from them. Some variables are ranked with an **interval scale** that does imply a specific distance (difference) between subjects. For example, college students are divided into freshmen, sophomore, etc. The 'distance' between them is, presumably, one year.

Numeric or **quantitative variables** measure the magnitude or quantity of the subjects' attributes. Numeric variables are usually divided into **continuous** and **discontinuous** variables. Continuous variables are, in principle, infinite, and allow any interval of numeral or fraction, depending on the accuracy of the measuring instrument. Common continuous numerical variables encountered by anthropologists are any size measure in archaeological remains: height, weight, etc. In practice, investigators working with continuous variables assign observations to an interval which contains several measurements. For example, if an anthropologist is measuring her subjects' heights, and one subject measures 156.113 cm and another 155.995 cm, she will probably assign both subjects to one category, namely, 156 cm. This topic will be discussed in more detail in section 2.3. White (1991) discusses the issue of measurement precision in osteological research. He specifically focuses on the appropriate procedure when slightly different measures of the same tooth or bone are obtained. The problems associated with measurement in osteology show that the measurement of continuous variables is approximate, and that perhaps the true value of a variate may be unknowable (Sokal and Rohlf, 1981).

Discontinuous numerical variables have specific values, with no intermediate values between them. For example, the number of household members, the number of cars in the household, or the number of children born to a woman can only be whole numbers. Of course, as we well know, these kinds of variables are amenable to statistical analysis, which sometimes produces counter-intuitive results. We have all heard reports of the average family size in the US as being 2.5 children per household, which simply means that the greater proportion of households has a family size close to that number. Please notice that it is possible to take the average of discrete numerical variables, whereas it is not possible to take the average of qualitative variables such as ethnic membership or status, even if they are coded with numbers.

1.2.6 Samples and statistics, populations and parameters; descriptive and inferential statistics; a few words about sampling

A **statistical population** is the entire group of individuals the researcher wants to study. Although statistical populations can be finite (all living children age 7 in one particular day) or infinite (all human beings when they were 7 years old), they tend to be incompletely observable (how could all children age 7 in the world be studied in one day?). A **parameter** is a measure (such as the average) that characterizes a population, and is denoted with Greek letters (for example, the population mean and standard deviation are designated with the Greek letters μ and σ, respectively). But as populations are usually incompletely observable, the value of the population parameter is usually unknown.

A **sample** is a subset of the population, and generally provides the data for research. A **statistic** is a measure that characterizes a sample. Thus, if a sample of children age 7 is obtained , its average height could easily be computed. Statistics are designated with Latin letters, such as $\bar{Y}$ (Y-bar) for the sample mean and s for the sample standard deviation. This difference in notation is very important because it provides clear information as to how the mean was obtained. It should be noted that population size is denoted with an uppercase N, whereas sample size is denoted with a lower case n.

How does a researcher know if a sample is representative of the population? There are excellent textbooks written on sampling procedures and on research design (Baker, 1994; Ellis, 1994; Judd, Smith and Kidder, 1991; Mascie-Taylor and Lasker, 1993). A discussion on sampling and research design is beyond the scope of this text. However, two issues should be discussed.

First, samples must be random. Imagine that an anthropologist is interested in studying the nutritional intake of children in a community. If the researcher only samples children from wealthy families, he is likely to have a non-representative impression of the actual nutritional intake of all children in the community. The conclusions of the study will be **biased**. A **represen-**

tative sample is usually defined as having being obtained through a procedure which gave every member of the population an equal chance of being sampled. This may be easier said than done in anthropology. An anthropologist in a particular community needs to understand the nuances and culture of the population, to make sure that an equal chance of being sampled was given to each and every member of the population. In many instances, common sense is the most important ingredient to a good sampling procedure.

There are some situations in anthropological research in which random sampling can hardly be attempted. For example, paleoanthropologists investigating populations of early hominids would hope to have a random sample of the entire population. But these researchers can only work with the animals that were fossilized. There is really no sampling procedure which could help them obtain a more representative sample than the existing fossil record. In this situation, the data are analyzed with the acknowledgment that they were obtained through a sampling procedure that can not be known to be random.

The second point to keep in mind is that samples should be of adequate size. Certainly, the larger the sample, the more similar it is to the entire population. But what exactly is large? This is not an easy question, especially because in anthropology it is sometimes impossible to increase a sample size. Paleoanthropologists keep hoping that more early hominids will be unearthed, but they can only work with what already exists. However, if a research project involves more easily accessible data sets, do notice that most statistical tests work well (are robust) with samples of at least 30 individuals. Indeed, there is a whole suit of non-parametric statistical tests specifically designed for (among other situations) cases in which the sample size is small.

Descriptive statistics describe the sample by summarizing raw data. They include measures of **central tendency** (the value around which much of the sample is distributed) and **dispersion** (how the sample is distributed around the central tendency value) such as the sample mean and standard deviation respectively. Descriptive statistics are of extreme importance if a research project does not lend itself to more complex statistical manipulations. A sample must be described accurately by providing clear numbers ('…11% of the sample does not believe in the supernatural…') instead of with words ('…few members of the sample do not believe in the supernatural…').

Inferential statistics are statistical techniques which use sample data, but make inferences about the population from which the sample was drawn. Most of this book is devoted to inferential statistics. Describing a sample is of essential importance, but scientists are interested in making statements about the entire population. Inferential statistics do precisely this.

1.3 Statistical notation

Variables are denoted with letters such as X, Y, and Z. In this book, if a single variable is discussed, it will be denoted with the letter Y. If more than one variable is measured in one individual, then we will differentiate them by using different letters. For example, Y could be the height, and X the weight of the subjects. Distinct observations can be differentiated through the use of subscripts. For example Y_1 is the observation recorded in the first individual, Y_2 in the observation recorded in the second one, and Y_n is the last observation, where n is the sample size.

Σ (sigma) is a summation sign which stands for the sum of the values that immediately follow it. For example, if Y stands for the variable height, and the sum of all the individuals' heights is desired, the operation can be denoted by writing ΣY. If only certain values should be added, say, the first ten, an index below sigma indicates the value at which summation will start, and an index above it indicates where summation will end. Thus:

$$\sum_{1}^{10} Y = Y_1 + Y_2 + \ldots + Y_{10}.$$

If it is clear that the summation is across all observations, no subscripts are needed. Below are two brief examples of the use of sigma (Σ):

X	Y
6	7
8	9
5	2
3	10
9	1
10	3
$\Sigma X = 41$	$\Sigma Y = 32$

A frequently used statistic is $(\Sigma Y)^2$, which is the square of the sum of the numbers. In our example:

$$(\Sigma X)^2 = (41)^2 = 1{,}681 \quad \text{and} \quad (\Sigma Y)^2 = (32)^2 = 1{,}024.$$

Another frequently used statistic is ΣY^2, which is the sum of the squared numbers. Using our previous examples, we can square the numbers, and sum them as follows:

X	X^2	Y	Y^2
6	36	7	49
8	64	9	81
5	25	2	4
3	9	10	100
9	81	1	1

X	X^2	Y	Y^2
10	100	3	9

$\Sigma X^2 = 315$ $\Sigma Y^2 = 244$

The reader should not confuse ΣY^2 with $(\Sigma Y)^2$. The former refers to the sum of squared numbers, whereas the latter refers to the square of the sum of the numbers. ΣY^2 is also commonly called the **uncorrected sum of squares** (USS). These two quantities are used in virtually every statistical test covered in this book.

If the sample includes two columns of numbers, it is possible to obtain additional statistics. If a column is multiplied by another, then a third column is obtained, which is the product of both columns. The sum of that third column is denoted by ΣYX. This is illustrated with the previous data:

X	Y	XY
6	7	42 (6×7)
8	9	72 (8×9)
5	2	10 (5×2)
3	10	30 (3×10)
9	1	9 (9×1)
10	3	30 (10×3)

$\Sigma YX = 193$ $(\Sigma XY)^2 = 37{,}249$

....

Practice problem 1.1

Obtain these statistics:

$$\Sigma X, \Sigma X^2, (\Sigma X)^2, \Sigma Y, \Sigma Y^2, (\Sigma Y)^2, \Sigma XY, (\Sigma XY)^2$$

using the following two columns of data

X	Y
0	15
70	80
100	6
10	50
20	75

Let us do the calculations step by step:

$\Sigma X = 0 + 70 + 100 + 10 + 20 = 200$

$\Sigma X^2 = (0)^2 + (70)^2 + (100)^2 + (10)^2 + (20)^2 = 0 + 4{,}900 + 10{,}000 + 100 + 400 = 15{,}400$

$(\Sigma X)^2 = (200)^2 = 40{,}000$

$\Sigma Y = 15 + 80 + 6 + 50 + 75 = 226$

$\Sigma Y^2 = (15)^2 + (80)^2 + (6)^2 + (50)^2 + (75)^2 = 225 + 6{,}400 + 36 + 2{,}500 + 5{,}625 = 14{,}786$

$(\Sigma Y)^2 = (226)^2 = 51{,}076$

$\Sigma XY = (0)(15) + (70)(80) + (100)(6) + (10)(50) + (20)(75) = 0 + 5,600 + 600 + 500 + 1,500 = 8,200$

$(\Sigma XY)^2 = (8,200)^2 = 67,240,000$

1.4 Rounding-off rules

The problem of rounding off arises when the number of digits in a particular figure should be reduced. Let us assume that the height of a subject is measured as 178.46 cm. If three digits are desired, then the rule that should be followed says that *the last significant digit is not rounded off if it is followed by a digit less than 5*. Thus, in this case the height would be reported as 178 because 8 is followed by 4. If four significant digits are desired, then the rule that should be followed says that *the last significant digit is increased by 1 if it is followed by a number greater than 5 or if it is followed by 5 followed by other non-zero digits*. In that case the height would be reported as being 178.5 because 4 is followed by 6. A slight problem arises when the figure is, for example, 178.5 or 177.50, and only three significant digits are desired. If the last digit was always increased or left unchanged, then the calculations would be artificially inflated or deflated. This problem is solved by using the following rule: *when the last significant digit is followed by a 5 standing alone or by a 5 followed by zeros, increase it if it is odd, and leave it unchanged if it is even*. In this manner, the number of figures which are rounded off by increasing the last significant digit will balance the number of figures whose last significant digit was left unchanged.

Practice problem 1.2

Round off the following figures to the stated number of significant digits:

Original figure	Number of significant digits	Rounded-off figure	Why
900.31	3	900	Last significant digit is left unchanged if followed by a number less than 5
900.7	3	901	Last significant digit is increased if followed by a number greater than 5 or by 5 followed by other non-zero numbers
901.5	3	902	If the last significant digit is followed by 5 alone or

Original figure	Number of significant digits	Rounded-off figure	Why
904.5	3	904	followed by zeros, it is increased if odd, but left unchanged if even.

1.5 Key concepts

Statistics
Theory
Hypothesis
Difference between scientific and non-scientific hypothesis
Difference between science and other forms of knowledge
Science and repeatability
The role of statistics in science
Validity and reliability
Variables and constants
Independent and dependent variables
Control and experimental groups
Randomization
Scales of measurement
Samples and statistics, populations and parameters
Descriptive and inferential statistics
Problems with sampling: randomness and size
Statistical notation and rounding rules

1.6 Exercises

1. Write down a research project in your own area of study. Answer the following.
   (a) Which variables are you going to measure, and are the variables quantitative (continuous or discontinuous), ranked, interval or qualitative?
   (b) Which variables are dependent and which independent?
   (c) Are you going to have control and experimental groups?
   (d) How are you going to obtain your sample?
   (e) How are you going to assure validity and reliability in your data?

2. Compute these statistics:

   ΣX, $(\Sigma X)^2$, ΣY, $(\Sigma Y)^2$, (ΣXY), $(\Sigma XY)^2$

   given the following data set.

X	Y
6	20
19	11
24	5
10	3
15	16
4	19
0	11
18	23
1	10

3. Compute these statistics:

$\Sigma X, \Sigma X^2, (\Sigma X)^2, \Sigma Y, \Sigma Y^2, (\Sigma Y)^2, (\Sigma XY), (\Sigma XY)^2$

given the following data set.

X	Y
30	100
25	98
10	105
16	67
34	77
20	101
34	100
35	45
19	80
55	107
68	90
55	108
38	78
70	100

4. Round the following figures to the specified number of digits.

Original number	Number of digits
100.32	3
100.78	3
100.50	3
99.50	3
178.897	5
198.214	5
201.555	5

2 | Frequency distributions and graphs

After a data set is collected, but before it is statistically analyzed, it is usually put in an easily understandable order. Frequency distributions and graphs are commonly used for this purpose. A graph presents, in an obvious manner, the shape or distribution of the data, and is a powerful visual aid in data presentation. A frequency distribution groups data into categories, and lists the number of observations which fall into such groups. In addition, the frequency distribution may contain more information such as a cumulative frequency column, etc.

There are two main reasons for constructing frequency distributions. The first is purely descriptive: the author wants to describe the data set. The second reason is as an aid to simplifying large data sets when statistical computations have to be done by hand. Thus, if a researcher has a large data set, and needs to compute statistics by hand, then she will probably group the data. With frequent use of computers today, more and more researchers are entering data sets into computers which do the calculations. The results of these computations using un-grouped data are more precise than those obtained with data grouped into frequency distributions. Because the former should be preferred over the latter, frequency distributions are infrequently used as a step in statistical calculations.

This chapter will review the construction of frequency distributions. It will also briefly discuss the use of graphs. With the availability of computers, it is now extremely easy to enter data, and obtain frequency distributions as well as graphs from the computer.

2.1 Frequency distributions of qualitative variables

In his ethnography of West Indians in Limon, Costa Rica, Purcell (1993) investigates the use of English by West Indians in an otherwise Spanish-speaking country. Of particular interest was the perception of English held by the West Indians in a society in which most business enterprises and all government institutions operate in Spanish. Thus, Purcell surveyed urban and rural subjects on their preference for Spanish or English. Table 2.1 shows a frequency distribution based on data presented in his table 6 (Purcell, 1993: 119).

Table 2.1. *Frequency distribution of a discrete variable*

Language	Data from rural survey				Data from urban survey			
	f	%	cf	cf%	f	%	cf	cf%
Spanish	17	19.54	17	19.54	65	50	65	50
English	65	74.71	82	94.25	65	50	130	100
No preference	5	5.75	87	100	—	—	—	—
	$n=87$	100%			$n=130$	100%		

Note:
cf: cumulative frequency
Source: Data from Purcell (1993)

The table as presented here includes more columns than are perhaps necessary in all circumstances. However, it illustrates how to construct a complete frequency distribution for qualitative variables.

The first step is to make sure that all categories are mutually exclusive and exhaustive. In other words, variates will fit in one and only one category.
In Purcell's study, there are three exhaustive, mutually exclusive categories: preference for Spanish, for English, and no preference. Notice that in the urban survey only two categories were needed to accommodate all subjects, whereas three were needed in the rural survey.

The second step in the construction of a frequency distribution is to enumerate the frequency (f or freq.) of individuals which fall into each category. This column adds up to the sample size (n). However, in this (and many other cases), a raw number is difficult to interpret, since the sample size of the urban survey ($n=87$) is different from the sample size of the rural survey ($n=130$). Hence, another column, the percentage (%) is also included. This is computed by dividing the frequency of a category by the sample size, and then multiplying by 100. Thus, the percentage for Spanish in the rural survey was obtained as (17/87)*100, for English as (65/87)*100, and for no preference as (5/87)*100.

The third step involves the last two columns, called the cumulative frequency (cf or cum. freq.) and the cumulative percentage (cf% or cum. freq. %). The former is basically a step for the computation of the latter. The cumulative frequency is obtained by adding to a frequency the frequency of the previous category. Thus, in the rural survey the frequency of preference for Spanish is 17, which is the first entry of the cf column. The second entry is computed by adding 17 to the frequency of the next category ($17 + 65 = 82$), and the last entry by adding the previous entry ($5 + 82 = 87$) to the last category's frequency. The last entry of cf should equal the sample size. The cumulative percentage is very informative, and is obtained by dividing the cf entries by sample size and multiplying it by 100. Thus, the first entry is

(17/87)*100 = 19.54. That means that 19.54% of all respondents fall into the category of preference for Spanish only. The next entry is obtained by dividing 82 by 87 and multiplying by 100 and is equal to 94.25. This means that 94.25% of subjects prefer Spanish or English. The last entry (no preference) adds up to 100. The last column is quite informative when comparing both urban and rural distributions: whereas in the rural survey 5.7% of subjects had no preference, 100% of subjects in the urban survey had a distinct preference for one or other language.

Frequency distributions are a descriptive tool. Therefore, the writer should decide what is necessary to include in a paper for the reader to have a good grasp of the sample. A researcher may wish to include less information than was covered here, but accompany the frequency distribution with a graph.

Practice problem 2.1

Below are the observed frequencies of hemoglobin phenotypes obtained by Madrigal (1989). Construct a complete frequency distribution.

Hemoglobin phenotypes	f	%	cf	cf%
AA	108	78.83	108	78.83
AS	22	16.06	130	94.89
SS	1	0.73	131	95.62
SC	2	1.46	133	97.08
AC	3	2.19	136	99.27
SF	1	0.73	$n=137$	100%
	$n=137$	100%		

First step: the categories are mutually exclusive and exhaustive. All variates are included, and they 'fit' into one and only one category.

Second step: the frequencies were given to you. Notice that the f column adds up to n. The percentage is computed as follows: Hb AA: (108/137)*100 = 78.83%, Hb AS: (22/137)*100 = 16.06%, Hb SS: (1/137)*100 = 0.73%, Hb SC: (2/137)*100 = 1.46%, Hb AC: (3/137)*100 = 2.19%, Hb SF(1/137)*100 = 0.73%.

Third step: The last two columns are computed. The cumulative frequency (cf) is simply obtained by writing down the first entry (108), then adding 22 (the next frequency) to 108: 103 + 22 = 130. The same is done for the subsequent entries: 130 + 1 = 131, 131 + 2 = 133, 133 + 3 = 136, and 136 + 1 = 137, which is our sample size. The last column is obtained by dividing the entries of the cf column by sample size, and multiplying by 100: Hb AA: (108/137)*100 = 78.83%, Hb AS: (130/137)*100 = 94.89%, Hb SS: (131/137)*100 = 95.62%, Hb SC: (133/137)*100 = 97.08%, Hb AC: (136/137)*100 = 99.27%, Hb SF(137/137)*100 = 100%. Notice that there was a great preponderance (94.89%) of the Hb AA and Hb AS phenotypes.

Table 2.2. *The number of births in a*
sample of 349 Shipibo females

Number of births	Females with that number of births
0	62
1	31
2	26
3	37
4	30
5	32
6	18
7	37
8	19
9	20
10	20
11	17
	$n=349$

Source: Data modified from Hern (1992a)

2.2 Frequency distributions of numerical discontinuous variables

The procedure for constructing frequency distributions for discontinuous numerical variables does not differ significantly from that of qualitative variables, unless the researcher decides to group several categories. Although discontinuous numeric variables can only take certain values, a researcher working with a large sample may wish to group the data by creating categories such as 0–2, 3–4, 5–7, more than 7 children, etc.

The data in table 2.2 were modified from a study by Hern (1992a) on the relation between polygyny and fertility among the Shipibo of the Peruvian Amazon. The data consist of the number of births recorded in females aged 15+. For the purposes of this book, we will only work with females who experienced from 0 to 11 births, excluding those who had 12 and 13. Table 2.2 lists (in the first column) the recorded number of births and (in the second column) the number of females who had that specific number of births.

Using these data, let us construct a frequency distribution as done in the previous section:

Number	(f)	%	cf	cf%
0	62	17.7	62	17.7
1	31	8.9	93	26.6
2	26	7.4	119	34.0
3	37	10.6	156	44.6
4	30	8.6	186	53.2
5	32	9.2	218	62.4

Number	(f)	%	cf	cf%
6	18	5.2	236	67.6
7	37	10.6	273	78.2
8	19	5.4	292	83.6
9	20	5.7	312	89.3
10	20	5.7	332	95.0
11	17	5.0	$n=349$	100.0

This shows that almost 18% of subjects did not experience a single term birth. The cf% column is particularly interesting, since it informs the reader that roughly 62% subjects had five or fewer births.

As the reader can see, nothing has been done differently from what was done when constructing frequency distributions for qualitative variables. But as mentioned before, if the number of categories in a frequency distribution is too large, or if some categories have frequencies of 0s, then the researcher may wish to group the data into larger categories (for example 0–2, 3–5, 6–8, etc.). These numbers are called the **limits of the interval**. Accordingly, if the data are grouped, the frequency distribution has to have a few extra columns, namely: the limits of the interval, and the **midpoint**, obtained as the difference between both limits divided by 2, and added to the lower limit. For example, the midpoint of the 0–2 interval is $(2-0)/2=1$, $1+0=1$. When grouping this kind of variable, it is best to use an odd number as a class interval so that the midpoint is a whole rather than a fractional number.

The data by Hern (1992a) presented above do not need to be grouped, as there are only 12 categories. However, let us practice grouping discontinuous numeric variables with these data. Let us agree that we will group them into intervals of 3 so that the midpoint is a whole number. Thus, our first interval will have limits of 0–2, with a midpoint of 1 ($(2-0)/2 = 1$; $1+0 = 1$), the second one will have limits of 3–5, with a midpoint of 4 ($(5-3)/2 = 1$; $1+3 = 4$), etc.

Limits	Midpoint	Number (f)	%	cf	cf%
0–2	1	119	34.1	119	34.1
3–5	4	99	28.4	218	62.5
6–8	7	74	21.2	292	83.7
9–11	10	57	16.3	$n=349$	100.0
		$n=349$	100%		

There are advantages and disadvantages to grouping a data set such as this. The advantage lies mostly in the computation of statistics: it is easier to compute, for example, the mean and the standard deviation with grouped data if the sample size is large. However, for purposes of description, the frequency distribution with the ungrouped data is better than the frequency distribution with the grouped categories, as long as the number of categories is not too large.

2.3 Frequency distributions of continuous numerical variables

Continuous numerical variables can take an infinite number of values. Thus, if the statures of two subjects are 156.0002 and 155.999 centimeters, they will probably be assigned to one category, namely 156. This will be done with the understanding that information is being lost, and that the final recorded measurement (156) is only an approximation to the reality of the subjects' heights. The process of constructing frequency distributions for continuous numerical data is only an extension of this grouping process. Thus, categories are going to have practical limits (also called apparent limits) that have the same level of precision as the original measure, and implied (or real) limits, which are half a unit more precise than the last significant unit originally used in our measurement. For example, if the original measures are (e.g.) 152, 153, 154, 155, 156, etc., the practical limits would be for example, 152–154, 155–157, etc., and the implied limits would be 151.5–154.5, 154.5–157.5, etc. The midpoint of the category is also computed, using the implied limits as follows: (upper implied limit − lower implied limit)/2 + lower implied limit. Therefore, if the implied limits are 157.5 and 154.5, then the midpoint would be calculated as $157.5 - 154.5 = 3$, $3/2 + 154.5 = 156$.

As always, the categories have to be mutually exclusive and exhaustive. There should also be enough categories to ensure that the data set is efficiently grouped. In other words, if the sample is large, say $n = 100$ and the values range from 1 to 90, and 90 of the observations have different values, the frequency distribution should not have 90 categories, as this hardly achieves any simplification. At the same time, if the data are grouped into two categories (1–44 and 45–90), most of the information is lost. Obviously, a compromise must be achieved. Different statistics textbooks have different guidelines for how many categories are desirable. Most cite between 10 and 20. However, the main issue to keep in mind is that the frequency distribution should not distort the distribution of the sample by grouping the variables into too few categories, but that efficiency and economy are achieved by not grouping them into too many categories. The researcher, who has the best knowledge of the data set, should make such decisions.

Table 2.3 shows part of a data set from Molnar et al. (1993: 138). The data were modified for didactic purposes by excluding the most extreme variate to facilitate grouping. Column c is the enamel cap area and column b is the combined area of dentin and pulp in 14 Neandertal mandibular molars (M1s). Scale is in square millimeters. Let us construct a frequency distribution with the first column.

Table 2.3

c	b
28.37	34.56
32.77	44.79
33.85	48.97
25.79	29.07
21.99	30.10
23.71	30.93
23.37	48.29
35.97	46.20
25.84	40.03
15.74	25.04
12.58	23.98
27.01	29.38
18.04	31.94

Source: Data modified from Molnar et al. (1993: 138)

First step: decide how many categories to include. The data range from a high value of 35.97 to a low value of 12.58, giving a span of 31.39 millimeters. Perhaps a good idea would be to start the first category right around 10 and finish over 35, with two categories per 10 units (5 units per category).

Second step: set up practical limits. The raw data have two decimal points. Since that is how precise the measurement was, that is how precise the practical limits will be. Thus, the first category could be started with 9.95 as its practical lower limit. The lower limit of the second category can be found by adding 5 units to 9.95, which adds up to 14.95.

To find the upper practical limit of each category, the same unit used for the lower practical limits is used. What is before 14.95, the low limit of the second category? 14.94, which is the upper limit of the first one. What's before 19.95, the lower limit of the third category? 19.94, which is the upper limit of the second one. Thus, the practical limits are:

9.95–14.94
14.95–19.94
19.95–24.94
24.95–29.94
29.95–34.94
34.95–39.94

Third step: find the implied limits. The implied limits extend half a unit beyond the last digit, and expand the category half a unit below the lower practical unit, and half a unit above the upper practical unit. If the reader is not sure how to find the first implied limit, I suggest that it is best to subtract from the lower practical limit 1/2 a unit. Therefore, from 9.95 subtract 0.005 (there were two decimal places in the original measurements). Thus, 9.95 − 0.005 = 9.945. From that, the other implied limits are obtained simply by adding 5 (the width of the interval). Thus, our lower implied limits are:

Practical limits	Lower implied limits
9.95–14.94	9.945
14.95–19.94	14.945
19.95–24.94	19.945
24.95–29.94	24.945
29.95–34.94	29.945
34.95–39.94	34.945

To find the upper implied limits, half a unit beyond the actual level of measurement is added to the upper practical limits. Therefore for the first category we obtain: 14.94 + 0.005 = 14.945. To the upper implied limits of the following categories simply keep adding 5, to obtain the following: 14.945 + 5 = 19.945. The complete implied limits are:

Practical limits	Implied limits
9.95–14.94	9.945–14.945
14.95–19.94	14.945–19.945
19.95–24.94	19.945–24.945
24.95–29.94	24.945–29.945
29.95–34.94	29.945–34.945
34.95–39.94	34.945–39.945

Fourth step: find the midpoint. The midpoint is obtained using the implied limits. For the first two categories, the midpoints are:

$$\frac{(14.945 - 9.945)}{2} + 9.945 = 12.445 \text{ for the first category, and}$$

$$\frac{(19.945 - 14.945)}{2} + 14.945 = 17.445 \text{ for the second one.}$$

Notice that 17.445 − 12.445 = 5, the width of the intervals. Thus, once the first midpoint is established, the width of the interval is added to it to obtain the second one, etc. The frequency distribution with the midpoints is:

Practical limits	Implied limits	Midpoints
9.95–14.94	9.945–14.945	12.445
14.95–19.94	14.945–19.945	17.445
19.95–24.94	19.945–24.945	22.445
24.95–29.94	24.945–29.945	27.445
29.95–34.94	29.945–34.945	32.445
34.95–39.94	34.945–39.945	37.445

Fifth step: construct the f, cf, % and cf% columns. This is done as previously. The entire frequency distribution is presented below.

Practical limits	Implied limits	Midpoints	f	cf	%	cf%
9.95–14.94	9.945–14.945	12.445	1	1	7.69	7.69
14.95–19.94	14.945–19.945	17.445	2	3	15.38	23.07
19.95–24.94	19.945–24.945	22.445	3	6	23.08	46.15
24.95–29.94	24.945–29.945	27.445	4	10	30.78	76.93
29.95–34.94	29.945–34.945	32.445	2	12	15.38	92.31
34.95–39.94	34.945–39.945	37.445	1	$n=13$	7.69	100.00

There is no set rule as to how to construct a frequency distribution. Thus, the practical limits could have been started at 10.95 instead of 9.95. The intervals could have been 4 instead of 5 units in width. What has been achieved by grouping the data? Instead of looking at 13 individual variates, the reader observes six intervals which show that the smallest and largest variates are the least frequent, and that the interval with the 27.445 midpoint is the most frequent. This conveys very well how these data are distributed. The cf% column also informs the reader that the first four intervals contain about 77 percent of all data. The frequency distribution conveys a better understanding of the data set. Let us construct a frequency distribution with the other column taken from Molnar et al. (column b).

Practice problem 2.2

Raw data for the enamel cap area (c) and the combined area of dentin and pulp (b) in 14 mandibular molars (M1s). Scale is in square millimeters. Data modified from Molnar et al. (1993). The variate with the highest value was removed to facilitate the grouping.

b
34.56
44.79
48.97
29.07
30.10
30.93
48.29
46.20
40.03
25.04
23.98
29.38
31.94

First step: decide how many categories to include. The data range from a high value of 48.97 to a low value of 23.98, giving a span of 24.99 millimeters. Perhaps a good idea would be to start the first category right around 20 and finish at 50, with two category per 10 units (5 units per category).

Second step: set up practical limits. The raw data have two decimal places. Thus, the practical limits will have as many.

Third step: find the implied limits. We need to extend the practical limits 1/2 unit beyond the last significant digit of the practical limits. Since the practical limits have two decimal places (.00), we subtract 1/2 a unit (.005) from the lower practical limit, and add 1/2 a unit to the upper practical unit (.005).

Fourth step: find the midpoint. The midpoints are computed using the implied limits. For the first category, the midpoint is

$$\left[\frac{25.005 - 20.005}{2} \right] + 20.005 = 22.505.$$

Again, notice that $27.505 - 22.505 = 5$, the width of our intervals. Thus, the other midpoints are obtained easily by adding 5 to the previous midpoint.

Fifth step: construct the f, cf, % and cf% columns. This is done as previously. The entire frequency distribution is presented below.

Practical limits	Implied limits	Midpoints	f	cf	%	cf%
20.01–25.00	20.005–25.005	22.505	1	1	7.69	7.60
25.01–30.00	25.005–30.005	27.505	3	4	23.08	30.77
30.01–35.00	30.005–35.005	32.505	3	7	23.08	53.85
35.01–40.00	35.005–40.005	37.505	1	8	7.69	61.54
40.01–45.00	40.005–45.005	42.505	2	10	15.38	76.92
45.01–50.00	45.005–50.005	47.505	3	$n = 13$	23.08	100.00

Frequency distribution using SAS/ASSIST

The partial Shipibo data shown above were used for producing this frequency distribution (Hern, 1992a). The data were entered into a SAS data set (a zero was entered 62 times, a one 31 times, etc). The menu-driven path was:

1. report writing,
2. counts,
3. lists,
4. one way tables.

The output is shown below:

Children produced

FERTILITY	Frequency	Percent	Cumulative Frequency	Cumulative Percent
0	62	17.8	62	17.8
1	31	8.9	93	26.6
2	26	7.4	119	34.1
3	37	10.6	156	44.7
4	30	8.6	186	53.3
5	32	9.2	218	62.5
6	18	5.2	236	67.6
7	37	10.6	273	78.2
8	19	5.4	292	83.7
9	20	5.7	312	89.4
10	20	5.7	332	95.1
11	17	4.9	349	100.0

2.4 Graphs

Graphs are a striking visual aid in communicating information. Their use, just like that of frequency distributions, is descriptive. Therefore, the researcher needs to decide which type of graph is most appropriate in what situation. Graphics production has become much simpler with the widespread use of computers. Indeed, a number of packages allow users to enter the data and graph it in different ways to decide which method of presentation is most appropriate for their needs. Because readers are likely to use different graphics packages, my discussion of graphs is limited, and includes: bar graphs, histograms, polygons and pie charts. In SAS/ASSIST, the menu-driven path is: 1. graphics, 2. high resolution, 3. a menu that offers the different kinds of graphs produced by SAS. Here the user would choose bar graphs, pie charts, etc. For histograms, however, the user must choose the menu-driven path for PROC CAPABILITY, with the histogram option.

2.4.1 Bar graphs

These graphs are used for qualitative as well as discontinuous numeric data, and show the frequency of different variates. The use of bar graphs is first illustrated with a qualitative variable, namely gender. I have collected historical demographic data from Escazu, a small rural parish in Costa Rica. Part of the data is marriage records from 1799 to 1900. Of interest was determining the gender of individuals who got married in the community, but whose surnames had not been previously recorded. If more males or more females had new surnames, it could be argued that surnames were differentially brought by male- or female-biased migration. In the decade of 1810–1819, 38 new names were recorded: 39 carried by males, 25 carried by females. The data are displayed in figure 2.1.

The graph demonstrates in a clear manner that more males than females brought new surnames into the community. Although this suggests that

Frequency

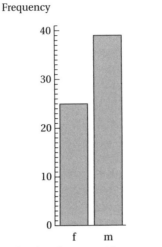

Gender of person with new name

Figure 2.1. A vertical bar graph of qualitative data. The number of males and females with new last names. Escazu, 1810–1819.

Frequency

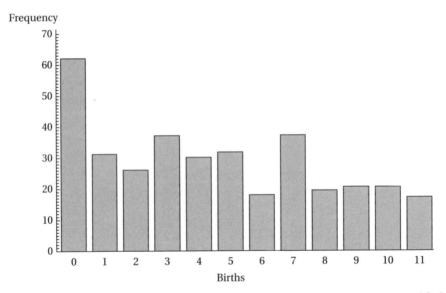

Births

Figure 2.2. A vertical bar graph of discontinuous quantitative data. (Data modified from Hern, 1992a.)

there may have been male-biased migration, we still have not conducted a statistical test to determine so.

The use of bar graphs is illustrated next with a discontinuous quantitative variable. Table 2.2 showed the number of children produced by a sample of 349 Shipibo females. The data are graphed in figure 2.2.

The graph shows the frequency distribution of the sample. SAS/ASSIST easily allows users to change the number of bars, in effect allowing them to group the data in any manner they wish.

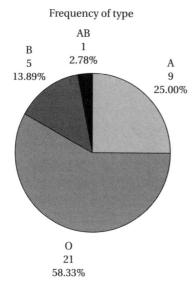

Frequency of type

Figure 2.3. A pie chart of a qualitative variable. Fictitious data on blood type distribution.

2.4.2 Pie charts

These charts are effective means of displaying qualitative or discontinuous numerical data. They clearly show what proportion of the total sample is composed by which variate. Let us say that the students of an introductory anthropology class are tested for their ABO blood type, and that their data are as given below:

Blood type	Frequency
O	21
A	9
B	5
AB	1
	$n = 36$

These frequencies could be displayed effectively with a pie chart (see figure 2.3).

2.4.3 Histograms

Histograms are used with continuous numerical data. They are similar to bar graphs, except that in histograms the bars are placed contiguous to each other, as they graph continuous data. In other words, whereas discontinuous numerical data have gaps between the variable's possible values, continuous numerical data have no gaps. Figure 2.4 shows a histogram of the dental data shown in table 2.3.

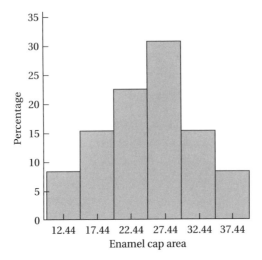

Figure 2.4. A histogram of continuous numerical data. (Data modified from Molnar et al., 1993.)

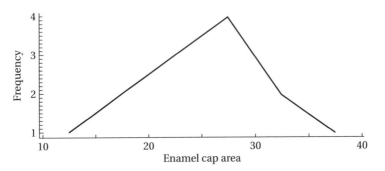

Figure 2.5. A polygon of continuous numerical data. Polygon of the frequency distribution of column c. (Data modified from Molnar et al., 1993.)

2.4.4 Polygons

These graphs are also applied to continuous numerical data, and are frequently preferred over histograms. Several computer packages refer to them as $X*Y$ graphs, in which X represents the variable, and Y the frequency of its various outcomes. Figure 2.5 shows a polygon of the same data plotted in the histogram (figure 2.4).

2.5 Key concepts

Frequency
Cumulative frequency percent (cf%)
Midpoint
Rounding error and use of computers
Appropriate categories and practical limits
Graphing discontinuous and continuous numerical data

2.6 Exercises

1. (a) Construct a frequency table from the following fictitious data. Age at late-onset (type II) Diabetes Mellitus in a sample of African American women:

   45 50 50 48 57 43 40 52 45 54 56 55 47 52 41 39

   (b) Construct categories with upper, lower and implied limits.
   (c) What is the midpoint of the ages? The mean?
   (d) Compute the cumulative frequency percentage. What does it tell you about the data?
   (e) What is/are the best graphing options for these data?

2. Below are the heights in centimeters of a fictitious sample of 19 adult males in a research project. Construct two frequency distributions, one with grouped and another one with un-grouped data. Graph both distributions.

187	190
199	187
193	176
167	188
189	192
179	183
190	180
189	194
189	185
197	

3. You conduct a survey on ethnic membership in a small community, and find the following results. Construct a frequency distribution, and graph your data with a pie chart.

OBS	Ethnicity of subjects
1	Chinese
2	Chinese
3	Chinese
4	Chinese
5	Chinese
6	Chinese
7	Chinese
8	Laotian
9	Laotian
10	Laotian
11	Laotian
12	Laotian

OBS	*Ethnicity of subjects*
13	Laotian
14	Laotian
15	Laotian
16	Laotian
17	Laotian
18	Laotian
19	Laotian
20	Laotian
21	Thai
22	Thai
23	Thai
24	Thai
25	Thai
26	Thai
27	Thai
28	Thai
29	Thai
30	Thai
31	Thai

$n = 31$

3 Descriptive statistics: measures of central tendency and dispersion

After constructing a frequency distribution, a researcher involved in quantitative analysis will usually compute sample descriptive statistics. Often, research papers do not include a frequency distribution, just descriptive statistics. Descriptive statistics convey two basic aspects of a sample: **central tendency and dispersion**. The former describes the most common variate of the sample, and the latter how the sample is distributed around the most common variate.

A word should be said about what calculation method may be easier in what situation. In a situation in which a researcher does not have access to a computer and needs to compute descriptive statistics of a large sample, he would probably choose to group the data into a frequency distribution because it is easier to compute the statistics in this manner. However, if the data are grouped, the results will not be as precise because of rounding error. With the use of computers, descriptive statistics are rarely computed by hand and therefore frequency distributions are rarely used for computation purposes. The computation of descriptive statistics with SAS/ASSIST will be illustrated at the end of the chapter, since all descriptive statistics can be obtained with one operation.

3.1 Measures of central tendency

This section discusses three central tendency statistics: the **mean**, the **median** and the **mode**. The three are different kinds of 'averages', used in different situations. Their general purpose is the same, namely, to find the single most representative score in the sample. Less commonly used measures of central tendency (such as the geometric and harmonic means) are covered by Sokal and Rohlf (1981).

3.1.1 The mean

This is the most commonly used statistic of central tendency for numeric continuous data. Indeed, when most people speak of 'the average', they are actually referring to the mean. The mean considers every observation, and is therefore a statistic affected by extreme variates. Thus, for samples with extreme values, the median (see below) is more appropriate. A very important property of the mean is that it is a non-biased estimate of the **population mean** (μ), and is the central-tendency measure of choice in most cases.

The **sample mean** is commonly represented by a capital Y capped by a bar ($\overline{Y}$), and referred to as 'Y-bar'. In textbooks which favor the use of X instead of Y, $\overline{X}$ is preferred instead. The mean is computed by summing all the observations in a sample, and dividing by the sample size. For example, if a student wants to obtain her mean test grade in a course in which three exams were administered (let's say her grades were 85, 90 and 100), the observations are summed ($85 + 90 + 100 = 275$), and divided by the sample size (3 in this case). Thus, the mean test score is $275/3 = 91.7$. The only difference between the population and sample mean formulae is that in the former N (the population size), and in the latter n (the sample size) is used. The formula for computing the sample mean is written as follows:

Formula 3.1 Formula for the sample mean

$$\overline{Y} = \frac{\Sigma Y}{n}$$

Let us compute the mean with the following data set:

5
6
8
7
10
100
8
9
6
10
10
5

First, add all the observations to obtain:

$\Sigma Y = 5 + 6 + 8 + 7 + 10 + 100 + 8 + 9 + 6 + 10 + 10 + 5 = 184$.

Then, count the number of observations to obtain your sample size (n). In our example, $n = 12$. Finally, divide ΣY by n. Thus $\overline{Y} = 184/12 = 15.3$.

This example illustrates well the point made above: that the mean can be influenced by extreme values. In this data set all but one observation were less than 11. However, because there was one observation whose value was 100, the mean is higher than 11, at 15.3.

Practice problem 3.1

The following data give the age of eight individuals attending a social service agency. Compute the mean age of the sample.

$$20 \quad 50 \quad 24 \quad 32 \quad 30 \quad 45 \quad 33 \quad 36$$

$\Sigma Y = 20 + 50 + 24 + 32 + 30 + 45 + 33 + 36 = 270$. Since $n = 8$, $\overline{Y} = 270/8 = 33.75$.

3.1.2 Computing the mean of frequency distributions

As mentioned before, if a researcher has no access to a computer, and wishes to compute the mean (as well as other descriptive statistics) of a large data set, he will probably want to group the data into a frequency distribution. If a frequency distribution is constructed with the sole purpose of computing descriptive statistics, then it does not need to be as complete as described in the previous chapter. Indeed, only the midpoint and the frequency of each category need to be included. Because of rounding error, the mean from data grouped in a frequency distribution will vary slightly from the mean computed with the un-grouped data. The following example illustrates this situation.

To compute the mean from grouped data, the frequency (f) of each category is multiplied by its midpoint (Y), and added across all categories ($\Sigma f Y$). This figure is then divided by the sample size.

Formula 3.2 Formula for the sample mean using grouped data

$$\overline{Y} = \frac{\Sigma f Y}{n}$$

Let us compute the mean of the following data set, first un-grouped and then grouped:

61 63 55 58 57 51 53 45 45 43 38 30 30 33 32 26 26
27 28 15 18 67 48 49 42 28 20 24 36 15

$\Sigma Y = 1163$, $n = 30$, $\overline{Y} = 38.77$

Y	Apparent limits	Real limits	Freq.	fY	
67	65–69	64.5–69.5	1	67	(1×67)
62	60–64	59.5–64.5	2	124	(2×62)
57	55–59	54.5–59.5	3	171	(3×57)
52	50–54	49.5–54.5	2	104	(2×52)
47	45–49	44.5–49.5	4	188	(4×47)
42	40–44	39.5–44.5	2	84	(2×42)

Y	Apparent limits	Real limits	Freq.	fY	
37	35–39	34.5–39.5	2	74	(2×37)
32	30–34	29.5–34.5	4	128	(4×32)
27	25–29	24.5–29.5	5	135	(5×27)
22	20–24	19.5–24.5	2	44	(2×22)
17	15–19	14.5–19.5	3	51	(3×17)
				$n=30$	

$$\Sigma fY = 67 + 124 + \cdots + 44 + 51 = 1{,}170, \quad n = 1 + 2 + \cdots + 2 + 3 = 30, \quad \overline{Y} = \frac{1{,}170}{30} = 39$$

This example illustrates that the means obtained by using grouped and ungrouped data are slightly different because detailed information is lost by grouping. However, this is deemed to be a small price to pay given the advantages of time saved by grouping the data into a frequency distribution. A last word about this example: sample size is obtained by adding up the frequencies. Beware of dividing ΣfY by the number of categories (in this case 11)!

Practice problem 3.2

Below is a frequency distribution showing the number of individuals living in 100 households. Compute the mean in the sample.

Y	f	fY
9	2	18
8	5	40
7	11	77
6	13	78
5	17	85
4	20	80
3	17	51
2	11	22
1	4	4

$$\Sigma fY = 18 + 40 + \cdots + 22 + 4 = 455, \quad n = 2 + 5 + \cdots + 11 + 4 = 100, \quad \overline{Y} = 455/100 = 4.55$$

3.1.3 The median

The median (commonly represented by a capital M) is another measure of central tendency used with numerical variables. It is typically used with data which do not follow a normal distribution (see next chapter). For now, suffice it to say that the median is commonly used in data sets with extreme values. The median is the score that divides a distribution exactly in half. Exactly one-half of the scores are less than or equal to the median, and exactly one-half are greater than or equal to it. This is the reason the median is not affected by extreme values. Readers are more likely to have seen reports of the median salary than the mean salary of a sample. For example, in a university there is a large spread of salaries, from the President of the university to graduate assistants. If a researcher were to obtain the mean salary at the university, she would be under the false impression that salaries are higher than they really are, all because of the very high salaries received by some members of the university. In this case, the investigator would be better off computing the median.

The median is also applied in open-ended distributions, in which the first or last category are of the kind '6 or more', or '1 or less'. These open-ended categories are frequent in surveys of family size, in which the researcher chooses to group in a category all individuals with extreme values. If the exact family size of some individuals, all of whom have been classified in a category named '6 or more', is unknown, the researcher would not be able to compute the mean family size in the sample.

The median is easily computed with a small sample size. With a large sample, the data need to be grouped into a frequency distribution. Both methods are discussed below.

If the data are not grouped, we follow these steps to compute the median:

1. Order the variates from lowest to highest.
2. If sample size is odd, the median is the center-most value. For example, in the data set 6 8 5 101 9, order it from lowest to highest as follows: 5 6 **8** 9 101. The median in this case is 8, the center-most value in the sample.
3. If sample size is even, the median is the mean of the two center-most values (the $\frac{n}{2}th$ and the $\frac{n}{2}th+1$). For example, for the data set 1 3 90 2 10 26 44 73, order it from lowest to highest as follows: 1 2 3 **10 26** 44 73 90. The median is the mean of the two center-most values, thus: $\frac{(10+26)}{2}=18$.

Practice problem 3.3

Compute the median of each of the following two sets of numbers.

1. 30 90 80 150 5 0 160. The first task is to order the data: 0 5 30 **80** 90 150 160. The median is the center-most value: 80.
2. 47 83 97 200 5 6. The first task is to order the data: 5 6 **47 83** 97 200. The median is the mean of the two center-most observations: $(47+83)/2=65$.

3.1.4 Computing the median of frequency distributions

The computation of the median of a moderately or large-size data set presents the difficulty of ordering the variates from lowest to highest without forgetting an observation. In this situation, it is best to construct a frequency distribution *including the real limits* to compute the median. Compute the median of a frequency distribution as follows:

1. Order the categories from lowest to highest.
2. Develop the frequency distribution including the real limits, the frequency of each category, and the cumulative frequency. As the categories are ordered from lowest to highest, start the cumulative frequency with the lower values.
3. Find the category that has the 50th **percentile** (P50) (see section 4.6). P50 is obtained by dividing sample size by 2. For example, if $n=150$, then $P50 = 150/2 = 75$. The interval which has P50 is the first one whose cumulative frequency exceeds the value of P50. Thus, if $P50 = 75$, and the cumulative frequency of two consecutive categories is 60 and 80, the latter one has P50.
4. Establish the width (W) of your category intervals.
5. In the formula to obtain the median, we will use the following notation:
    $n=$ the sample size
    $l_p=$ the lower real limit of the interval containing P50
    $c_p=$ the cumulative frequency up to but not including the interval containing P50
    $f_p=$ the frequency of the interval containing P50
    $W=$ the width of the intervals in your frequency distribution.
6. The formula for computing the median with a frequency distribution looks more cumbersome than it really is:

Formula 3.3 Formula for the median using grouped data

$$M = l_p + \left[\frac{(0.5 \times n) - c_p}{f_p} \right] W$$

There is a definite order in which these operations need to be performed:
   (a) Multiply the sample size by 0.5.
   (b) From that product subtract c_p.
   (c) Divide that number by f_p.
   (d) Multiply the quotient of that division by the width of the intervals.
   (e) Add l_p to that product.

Let us compute the median of the following frequency distribution. Included are the real limits, as well as the frequency and cumulative frequency columns.

Y	Real limits	f	cf
1	0.50–1.50	1	1
2	1.50–2.50	3	4
3	2.50–3.50	8	12
4	3.50–4.50	6	18*

Y	Real limits	f	cf
5	4.50–5.50	4	22
6	5.50–6.50	2	24
7	6.50–7.50	4	28
8	7.50–8.50	2	30 = n

1. The categories have been ordered from lowest (1) to highest (8).
2. The cumulative frequency column has been started with the lowest values.
3. P50 is obtained by dividing sample size by 2. Thus, in this case it is 30/2 = 15. The interval which has P50 is the first one whose cumulative frequency exceeds the value of P50. Thus, in this example the category whose cumulative frequency is 18 has P50 (the cumulative frequency before it is 12, which does not exceed 15). The interval with the P50 has been marked with an asterisk (*).
4. The width (W) of the category intervals is 1.
5. The following numbers are established:
    $l_p = 3.5$
    $c_p = 12$
    $f_p = 6$
    $W = 1$
6. The median is computed as follows:

$$M = 3.5 + \left[\frac{(0.5)(30) - 12}{6} \right](1) = 3.5 + \left[\frac{15 - 12}{6} \right](1) = 3.5 + \left[\frac{3}{6} \right] = 3.5 + [0.5] \, (1)$$
$$= 3.5 + 0.5 = 4$$

It has been mentioned repeatedly that the frequency distribution has to be ordered from the lowest to the highest category. If a frequency distribution is ordered from highest to lowest, the median can be computed using the formula above, *as long as the cumulative frequency is started from the lowest to highest categories.* Let us compute the median with the previous data set, but arranged differently.

Y	Real limits	f	cf
8	7.50–8.50	2	30
7	6.50–7.50	4	28
6	5.50–6.50	2	24
5	4.50–5.50	4	22
4	3.50–4.50	6	18*
3	2.50–3.50	8	12
2	1.50–2.50	3	4
1	0.50–1.50	1	1

1. The categories have been ordered from highest (8) to lowest (1).
2. The cumulative frequency column has been started with the lowest values, namely the last category.
3. P50 is 30/2 = 15. The interval which has P50 is the one whose cumulative frequency is 18 (the first to exceed P50).

4. The width (W) of the category intervals is 1.

5. The following numbers are established:

$l_p = 3.5$

$c_p = 12$

$f_p = 6$

$W = 1$

6. The median is computed:

$$M = 3.5 + \left[\frac{(0.5)(30) - 12}{6} \right](1) = 3.5 + \left[\frac{15 - 12}{6} \right](1) = 3.5 + \left[\frac{3}{6} \right] = 3.5 + [0.5](1)$$

$$= 3.5 + 0.5 = 4$$

Practice problem 3.4

The following data are the yearly salaries (in US dollars) of a sample of university employees. Compute the median salary in the sample.

Midpoint	Apparent limits	Real limits	f	cf
10,000	0–19,999.00	–0.5–19,999.50	30	30
30,000	20,000.00–39,999.00	19,999.50–39,999.50	20	50
50,000	40,000.00–59,999.00	39,999.50–59,999.50	15	65*
70,000	60,000.00–79,999.00	59,999.50–79,999.50	12	77
90,000	80,000.00–99,999.00	79,999.50–99,999.50	11	88
110,000	100,000.00–119,999.00	99,999.50–119,999.50	10	98
130,000	120,000.00–139,999.00	119,999.50–139,999.50	8	106
150,000	140,000.00–159,999.00	139,999.50–159,999.50	5	111
170,000	160,000.00–179,999.00	159,999.50–179,999.50	3	114
190,000	180,000.00–199,999.00	179,999.50–199,999.50	1	$n = 115$

Let us do this problem step by step:

1. The categories have been ordered from lowest (10,000) to highest (190,000) .

2. The cumulative frequency column has been started with the lowest values.

3. P50 is $\frac{115}{2} = 57.2$. The interval which has P50 is the first one whose cumulative frequency exceeds the value of P50. Thus, in this example the category whose cumulative frequency is 65 has P50. The interval with the P50 has been marked with an asterisk (*).

4. The width (W) of the category intervals is 20,000. This was obtained by subtracting say, 10,000 from 30,000 (the midpoints of the first two categories).

5. The following numbers are established:

$l_p = 39,999.5$

$c_p = 50$

$f_p = 15$

$W = 20,000$

6. The median is computed:

$$M = 39{,}999.5 + \left[\frac{(0.5 \times 115) - 50}{15}\right](20{,}000) = 39{,}999.5 + \left[\frac{7.5}{15}\right](20{,}000)$$

$$= 39{,}999.5 + 10{,}000 = 49{,}999$$

This value is lower than the mean salary, which is $61,478.26. This example clearly shows that, whereas the mean is influenced by extreme values, the median is not.

3.1.5 The mode

Archaeologists are frequently confronted with the task of classifying pottery according to types of decoration or even color. Let us presume that a student has been hired by an archaeologist to classify the remains of 50 pots, and to write a short summary of his findings. The assistant assigns a code to each coloration, and constructs the following frequency distribution:

Type of paint	Code	Freq.
Plain	1	30
Black	2	15
Red	3	5

Which one is the average type? In this situation, the mode, or the most common value, should be reported. As a matter of fact, any time that researchers deal with qualitative variables (gender, ethnic group, blood type, etc.), they should use the mode to report central tendency. In the example above, the modal ceramic type is the 'plain' one. In some distributions, two categories have virtually equal frequencies. Such distributions are called bimodal, and have two modes.

The mode can also be obtained for a frequency distribution of numerical variables. The mode is easily determined by examining the frequency distribution, and finding the category with the highest frequency. Let us determine the mode in the frequency distribution below:

Y	9	8	7	6	5	4	3	2	1
f	2	5	6	1	17	20	4	1	4

The mode in this example is 4, which has a frequency of 20.

Practice problem 3.5

The following data are the yearly salaries of a sample of university employees. What is the modal salary in the sample?

Midpoint	Apparent limits	Real limits	f	cf
10,000	0–19,999.00	–0.5–19,999.50	30	30
30,000	20,000.00–39,999.00	19,999.50–39,999.50	20	50
50,000	40,000.00–59,999.00	39,999.50–59,999.50	15	65
70,000	60,000.00–79,999.00	59,999.50–79,999.50	12	77
90,000	80,000.00–99,999.00	79,999.50–99,999.50	11	88
110,000	100,000.00–119,999.00	99,999.50–119,999.50	10	98
130,000	120,000.00–139,999.00	119,999.50–139,999.50	8	106
150,000	140,000.00–159,999.00	139,999.50–159,999.50	5	111
170,000	160,000.00–179,999.00	159,999.50–179,999.50	3	114
190,000	180,000.00–199,999.00	179,999.50–199,999.50	1	$n=115$

The mode in this case is the category with the lowest salary, whose frequency is 30. This example confirms that, even though the median is $M=\$49,999.5$, and the mean is $\$61,478.26$, the most common salary in the sample (and probably in the university community) is the lowest one.

3.2 Measures of variation

As mentioned previously, descriptive statistics summarize the sample by conveying the average observation, as well as the variation around it. A good measure of variability should provide an accurate picture of the spread of the distribution; in other words, it should give an indication of how accurately the average (whichever is chosen) describes the distribution. If the variability is small, then the scores are close to the average. When variability is large, the scores are spread out, and are not necessarily close to the average.

3.2.1 The range

The **range** is simply the difference between the highest and lowest variates plus one. The one is added to include the real limits of the data. Although other statistics texts choose to define the range as the difference between the highest and lowest value without adding the one (Bohrnstedt and Knoke, 1988; Sokal and Rohlf, 1981), here we will add the one, to agree with SAS' computations. The range is used primarily when the researcher wants to convey how much difference there is between the lowest and highest score. I have found it useful when describing the spread of variation of live births in a sample of non-contracepting women: in my study the range was a whopping $17-0+1=18$ births (Madrigal, 1989). However, as the range only takes into account two values, it does not express variation within the sample very well. By knowing that the range in my sample was 18, the reader still does not know how the fertility varies around the mean.

Compute the range for the number of cows owned by six families.

10 7 5 5 2 1

The range is $10 - 1 + 1 = 10$

3.2.2 The population variance and standard deviation; the definitional formulae

The reader may recall from chapter 1 that it is desirable for sample statistics to be non-biased estimators of population parameters. If sample statistics are biased they do not provide a good approximation of the true value of the population parameter. Whereas the sample mean is an unbiased estimator of the population mean, the sample standard deviation and variance are biased estimators of the population standard deviation and variance: they tend to underestimate the value of the parameters. Therefore, the population formula can not be applied to a sample when computing the sample variance and standard deviation. This section covers the computation of the population variance and standard deviation, and the next the computation of the sample variance and standard deviation.

Just like the mean, the population and sample variance and standard deviation have different symbols: The notation for the **sample variance** and the **standard deviation** is s^2 and s respectively. The notation for the **population variance** and **standard deviation** is σ^2 and σ respectively.

First, a few words about the variance and the standard deviation. The great advantage they have over the range is that they take into account all values in the population instead of the two extremes. Therefore, they provide a much better idea of the amount of variation around the mean. Let us take as a didactic example the fertility data previously noted. The mean number of live births is 5.97, with a range of 18 (Madrigal, 1989). How is all the sample (not just the highest and lowest observations) distributed around the mean? If the difference between each woman's observation and the sample mean is computed, it can be established how far away each woman's datum is from the mean. But what is desirable is a measure that summarizes the difference between the fertility of all women and their mean fertility. The standard deviation provides that information: it considers the distance between each score and the mean. The variance has a very similar 'meaning'. The population variance is the mean *squared* deviation from the mean. In other words, the variance computes the difference between each observation and the mean, squares it, and averages all the differences. Because the variance works with the squared distances from the mean, it is a large number relative to the original scores. To obtain a measure back in the original scale of measurement, the square root of the variance is computed. The standard deviation is the square root of the variance.

Of the two statistics, the standard deviation is more commonly used. Below are the formulae for the population variance and standard deviation. However, the sum of squares (SS), also known as the corrected sum of squares (CSS), is first introduced. As its name implies, it is the sum of the squared differences between the observations and their mean.

Formula 3.4 Formula for the population sum of squares (SS)

$$SS = \Sigma(Y - \mu)^2$$

Formula 3.5 Formula for the population variance

$$\sigma^2 = \frac{\Sigma(Y - \mu)^2}{N}$$

Formula 3.6 Formula for the population standard deviation

$$\sigma = \sqrt{\sigma^2}$$

Let us assume that the following is a population, and we wish to compute its variance and standard deviation.

Y
10
7
5
5
2
1

Let us follow these steps:
1. First, compute the mean: $\mu = 5$.
2. Second, obtain the difference between the mean and each observation.

Y	$(Y - \mu)$
10	$10 - 5 = 5$
7	$7 - 5 = 2$
5	$5 - 5 = 0$
5	$5 - 5 = 0$
2	$2 - 5 = -3$
1	$1 - 5 = -4$

3. Add the differences between the observations and the mean: $5 + 2 + 0 + 0 + -3 + -3 = 0$. If a population's or sample's mean is subtracted

from each observation, and the differences are added, the result will be zero. This is why the differences must be squared.

4. Square the differences and add them up to obtain the sum of squares.

Y	$(Y-\mu)$	$(Y-\mu)^2$
10	$10-5=5$	25
7	$7-5=2$	4
5	$5-5=0$	0
5	$5-5=0$	0
2	$2-5=-3$	99
1	$1-5=-4$	16

Thus, $SS=\Sigma(Y-\mu)^2=54$.

5. Divide SS by N to obtain the population variance $\sigma^2=54/6=9$.

6. To obtain the population standard deviation, compute the square root of the variance: $\sigma=\sqrt{9}=3$.

Practice problem 3.7

Compute the variance and standard deviation of the following population.

Y	$(Y-\mu)$	$(Y-\mu)^2$
30	-4.67	21.8
33	-1.67	2.8
41	6.33	40.1
43	8.33	69.4
29	-5.67	32.1
32	-2.67	7.1

1. Compute the mean: $\mu=34.67$.

2–4. Compute the sum of squares: $SS=\Sigma(Y-\mu)^2=173.33$.

5. Divide the sum of squares by sample size ($N=6$) to obtain the variance:

$$\sigma^2=173.33/6=28.88$$

6. Take the square root of the variance to obtain the standard deviation:

$$\sigma=\sqrt{\sigma^2}=\sqrt{28.88}=5.37$$

3.2.3 The sample variance and standard deviation; the definitional formulae

If the sample variance and standard deviations were computed with the population formulae, they would provide a biased estimator of the true parameters. Fortunately, the difference between the sample and population formulae is quite small: the sum of squares is divided by $n-1$ instead of by n.

The quantity $n-1$ is known as the degrees of freedom, and will be used repeatedly later in the book.

Below are the formulae for the computation of the sample variance and standard deviation. The parameter notation is no longer used.

Formula 3.7 Formula for the sample variance

$$s^2 = \frac{\Sigma(Y-\overline{Y})^2}{n-1}$$

Formula 3.8 Formula for the sample standard deviation

$$s^2 = \sqrt{s^2}$$

To illustrate the difference between the population and sample formulae, let us compute the variance and standard deviation with the data set we used previously to compute the population parameters.

Y
10
7
5
5
2
1

Let us follow these steps:
1. First, compute the mean: $\overline{Y}=5$.
2. Second, obtain the difference between the mean and each observation.

Y	$(Y-\overline{Y})$
10	$10-5=5$
7	$7-5=2$
5	$5-5=0$
5	$5-5=0$
2	$2-5=-3$
1	$1-5=-4$

3. Square the differences and add them up to obtain the SS.

Y	$(Y-\overline{Y})$	$(Y-\overline{Y})^2$
10	$10-5=5$	25
7	$7-5=2$	4
5	$5-5=0$	0
5	$5-5=0$	0
2	$2-5=-3$	9
1	$1-5=-4$	16

Thus, $\Sigma(Y-\overline{Y})^2 = 54$.

4. Divide $\Sigma(Y-\overline{Y})^2$ by $n-1$ to obtain the sample variance: $s^2 = 54/5 = 10.8$. Notice that the variance we computed with the population formula was lower, namely 9.

5. To compute the population standard deviation, obtain the square root of the variance: $s = \sqrt{10.8} = 3.3$. As you can see, the standard deviation obtained with the sample formula is also higher than that computed with the population formula ($\sigma = 3$). Clearly then, the population formula will produce values which are lower than those obtained by applying the sample formula.

Practice problem 3.8

Compute the variance and standard deviation of the following sample of ages of clients of a social service agency.

Y	$(Y-\overline{Y})$	$(Y-\overline{Y})^2$
30	−4.67	21.8
33	−1.67	2.8
41	6.33	40.1
43	8.33	69.4
29	−5.67	32.1
32	−2.67	7.1

1. Compute the mean $\overline{Y} = 34.67$.

2,3. Compute the sum of squares: $\Sigma(Y-\overline{Y})^2 = 173.33$.

4. Divide the sum of squares by sample size minus 1 ($n = 6$) to obtain the variance:

$$s^2 = 173.33/5 = 34.666$$

5. Take the square root of the variance to obtain the standard deviation:

$$s = \sqrt{s_9} = \sqrt{34.666} = 5.9$$

Compare these results with those obtained in practice problem 3.7.

3.2.4 The population and sample variance and standard deviation; the computational ('machine') formula

The population and sample formulae for the variance and standard deviation presented in the previous sections are called the definitional formulae because they define mathematically what these statistics are. With these formulae, we computed the difference between observations and their mean, squared them, added them and divided their sum by the population size or $n-1$ to obtain the variance. Then we took the square root of the variance to compute the standard deviation. However, the definitional formulae are cumbersome and can easily accumulate rounding errors associated with adding squared differences. Therefore, it is better to use other formulae (called the computational or machine formulae) which are easier to work with, especially when working with large sample sizes and with a calculator. The standard deviation is computed as before, by taking the square root of the variance (see formulae 3.6 and 3.8).

Formula 3.9 Formula for the sum of squares (SS). 'Machine' computation (use n if working with a sample, N if working with a population)

$$SS = \Sigma Y^2 - \frac{(\Sigma Y)^2}{n}$$

Formula 3.10 Formula for the population variance. 'Machine' computation

$$\sigma^2 = \frac{\Sigma Y^2 - \frac{(\Sigma Y)^2}{N}}{N}$$

Formula 3.11 Formula for the sample variance. 'Machine' computation

$$s^2 = \frac{\Sigma Y^2 - \frac{(\Sigma Y)^2}{n}}{n-1}$$

First, let us remind ourselves of what these statistics are: ΣY is the sum of the observations, $(\Sigma Y)^2$ is the sum of the observations squared, and ΣY^2 is the sum of the squared observations. These operations must be performed in a specific order. First, square ΣY to obtain $(\Sigma Y)^2$. Then divide $(\Sigma Y)^2$ by sample (or population) size. Then, the quotient of the latter division should be subtracted from ΣY^2. Finally, perform the last division by $n-1$ if working with a sample or N if working with a population.

Therefore, the difference between the sample and population variance is still that the sum of squares is divided in one case by $n-1$, and by N in the other. To compute the standard deviation, the square root of the variance is taken (see formulae 3.6 and 3.8). The big difference between the machine and definitional formulae is the manner in which the sum of squares is computed. Let us use the data sets we used, applying *the sample formula*. For the reader's convenience, a column of the numbers squared is also included:

Y	Y^2
10	100
7	49
5	25
5	25
2	4
1	1

Let us follow these steps:

1. Compute ΣY and square it: $\Sigma Y = 30$, and $(\Sigma Y)^2 = 900$.
2. Square the numbers and add them up to obtain ΣY^2. In this case, $\Sigma Y^2 = 204$.
3. The sample size is $n = 6$, and the degrees of freedom $n - 1 = 5$.
4. Compute the variance:

$$s^2 = \frac{204 - \dfrac{(30)^2}{6}}{5} = \frac{204 - \dfrac{900}{6}}{5} = \frac{204 - 150}{5} = \frac{54}{5} = 10.8$$

5. The standard deviation is the square root of the variance:

$$s = \sqrt{10.8} = 3.3.$$

The values of the sum of squares, the variance and the standard deviation are all the same as those computed with the definitional formula.

Practice problem 3.9

Compute the variance and standard deviation of the following sample of ages of clients attending a social service agency.

Y	Y²
30	900
33	1,089
41	1,681
43	1,849
29	841
32	1,024

1. Compute ΣY and square it: $\Sigma Y = 208$, and $(\Sigma Y)^2 = 43{,}264$.

2. Square the numbers and add them up to obtain ΣY^2. In this case, $\Sigma Y^2 = 7{,}384$.

3. Keep in mind that $n = 6$, and $n - 1 = 5$.

4. Compute the variance:

$$s^2 = \frac{7{,}384 - \dfrac{(208)^2}{6}}{5} = \frac{7{,}384 - \dfrac{43{,}264}{6}}{5} = \frac{7{,}384 - 7{,}210.67}{5} = \frac{173.33}{5} = 34.7$$

5. The standard deviation is the square root of the variance:

$$s\sqrt{34.67} = 5.9$$

3.2.5 The computational ('machine') formula with frequency distributions

If a sample is relatively large, it is easier to compute the variance and the standard deviation with the data grouped into a frequency distribution in which only the midpoint and frequency columns are strictly necessary. The machine formula remains virtually the same, with the same difference between the population and sample formulae: the sum of squares is divided by N if working with a population, and by $n-1$ if working with a sample. Below is the formula for computing the population and sample variance. The standard deviation is computed as before, by taking the square root of the variance (see formulae 3.6 and 3.8).

Formula 3.12 Formula for the sum of squares (SS). 'Machine' computation. Grouped data (use n if working with sample, N if working with population)

$$SS = \Sigma fY^2 - \frac{(\Sigma fY)^2}{n}$$

Formula 3.13 Formula for the population variance. 'Machine' computation. Grouped data

$$\sigma^2 = \frac{\Sigma fY_9 - \dfrac{(\Sigma fY)_9}{N}}{N}$$

Formula 3.14 Formula for the sample variance. 'Machine' computation. Grouped data

$$s^2 = \frac{\Sigma fY_9 - \dfrac{(\Sigma fY)_9}{n}}{n-1}$$

Formulae 3.12–3.14 are different from formulae 3.9–3.11 only in that the former include the 'f' for frequency. Thus, instead of simply adding up the numbers to obtain ΣY, we now multiply each midpoint by its frequency, and add up the products to obtain ΣfY. The same goes for ΣfY^2. First, the midpoints are squared, then multiplied by their frequency, and summed. The following example takes the reader step by step through the computation of the *sample* variance and standard deviation with a frequency distribution. How these columns were computed is illustrated with the first row.

Y	f	fY	Y²	fY²
8	2	16 (8×2)	64 (8×8)	128 (2×64)
7	4	28	49	196
6	12	72	36	432
5	22	110	25	550
4	24	96	16	384
3	19	57	9	171
2	15	30	4	60
1	9	9	1	9

Let us go through the following steps.

1. Compute the necessary columns fY, Y^2 and fY^2 and compute ΣfY and ΣfY^2. In this case $\Sigma fY = 418$ and $\Sigma fY^2 = 1{,}930$.

2. Sample size in this case is $n = 107$.
3. Compute the variance:

$$s = \frac{1{,}930 - \frac{(418)_9}{107}}{106} = \frac{1{,}930 - \frac{174{,}724}{107}}{106} = \frac{1{,}930 - 1{,}632.93}{106} = \frac{297.06}{106} = 2.80$$

4. The standard deviation is obtained by taking the square root of the variance.

$$s = \sqrt{2.80} = 1.67$$

Practice problem 3.10

The following data set shows the weight in pounds of a group of 25 children aged 4 years. Compute the variance and standard deviation of the sample.

Y	f	fY	Y^2	fY^2
34	2	68	1,156	2,312
36	8	288	1,296	1,0368
38	5	190	1,444	7,220
40	8	320	1,600	1,2800
42	2	84	1,764	3,528

1,2. Compute the following statistics: $n = 25$, $\Sigma fY = 950$, $\Sigma fY^2 = 36{,}228$.
3. Compute the variance:

$$s = \frac{36{,}228 - \frac{(950)_9}{25}}{24} = \frac{36{,}228 - \frac{902{,}500}{25}}{24} = \frac{36{,}228 - 36{,}100}{24} = \frac{128}{24} = 5.33$$

4. The standard deviation is the square root of the variance:

$$s = \sqrt{s^2} = \sqrt{5.33} = 2.31$$

Descriptive statistics using SAS/ASSIST

The data for this practice consist of oyster shell metric measurements from the Van Horn Creek site (data kindly provided by White, work in progress), and include the weight in grams of individual oyster 'bottom' valves. Only part of the data is shown.

112.1
74.1
68.9
35.2
55.0
21.2
61.5
9.1
40.9
26.1
30.0

SAS has two procedures that compute descriptive statistics: PROC MEANS and PROC UNI-VARIATE. However, the latter one is not available through the menus, and has to be accessed via the program editor. PROC MEANS is easily accessed by following this path: 1. data analysis, 2. elementary, 3. summary statistics, 4. at the summary statistics menu, the user chooses what statistics to compute. The following output is obtained:

```
N     Nmiss    Minimum      Maximum       Range        Sum           Mean
----------------------------------------------------------------------------
11    0        9.1000000    112.1000000   103.0000000  534.7000000   48.6090909

                          Variance      Std Dev
                          -----------------------
                          873.4989091   29.5550150
                          -----------------------
```

To run PROC UNIVARIATE the user can write the following code in the program editor

```
options linesize = 80 pagesize = 54 date number pageno = 1;
title;
footnote;
proc univariate data = BOOK.NONPAR ; var weight ;
run;
```

PROC UNIVARIATE provides a more in-depth analysis of the data. Some of the statistics computed by it, and not by PROC MEANS, are marked. The output is:

```
                        Univariate Procedure
Variable = WEIGHT
                             Moments
            N                 11    Sum Wgts        11
            Mean         48.60909   Sum            534.7
            Std Dev      29.55501   Variance     873.4989
            Skewness     0.859157   Kurtosis     0.717117
  ΣY²→      USS          34726.27   CSS          8734.989    ←SS
            CV           60.80141   Std Mean     8.911172
            T:Mean = 0   5.454848   Pr>|T|         0.0003
            Num ^ = 0         11    Num>0           11
            M(Sign)          5.5    Pr>=|M|        0.0010
            Sgn Rank          33    Pr>=|S|        0.0010

                          Quantiles (Def = 5)
            100% Max       112.1    99%            112.1
             75% Q3         68.9    95%            112.1
             50% Med        40.9    90%             74.7
             25% Q1         26.1    10%             21.2
              0% Min         9.1    5%               9.1
                                    1%               9.1
            Range            103
            Q3–Q1           42.8
            Mode             9.1

                             Extremes
            Lowest       Obs   Highest       Obs
             9.1(         8)     55(          5)
            21.2(         6)     61.5(        7)
            26.1(        10)     68.9(        3)
              30(        11)     74.7(        2)
            35.2(         4)    112.1(        1)
```

Table 3.1. *Household composition and size according to ranked sample communities*

Ranked communities	Mean size	Standard deviation	Percentage grand-parents in community	Percentage grand-children in community	Percentage H[a] members not HH lineal issue
U1 (N=28)	5.2	2.71	6.1	9.6	12.9
U2 (N=37)	5.4	2.27	<1.0	3.3	5.8
U3 (N=32)	4.1	2.44	<1.0	7.9	7.4
R1 (N=47)	4.3	2.86	4.1	4.1	7.3
R2 (N=40)	4.3	2.48	<1.0	4.3	5.5

Notes:
[a]H household; HH household head
Source: From Purcell (1993:145). Reprinted with permission from the publishers.

3.3 A research example of descriptive statistics

In his study of West Indians in Limon Costa Rica, Purcell (1993) investigated the household composition in rural and urban areas of Limon in relation to migration. According to him, the availability of economic opportunities affects household composition and size, by attracting people of specific ages to specific communities. He ranks three urban (U1–U3) and two rural communities (R1–R2) according to the relative distribution of occupations in each (teaching, clerical, commercial, farming, etc.). In the table reproduced here as table 3.1, he reports the mean and standard deviation to summarize the size and variation of size of households in these communities. The table shows that the urban communities have larger mean households, with means of 5.2 and 5.4, as opposed to means of 4.1 and 4.3 (for two) rural sites. His table also includes information (in the form of percentages) on the household membership of different communities. The communities differ in the percentage of grandparents or grandchildren living in the community, and in the percentage of household members who are not part of the household head's lineal family. According to Purcell (1993), 'the inescapable conclusion is that migration, usually an aspect of social mobility, plays a significant part in household size and composition'.

3.4 Key concepts

The importance of describing a sample, instead of providing raw data
Measures of central tendency and variation
The difference between the population and sample variance and standard
     deviation formulae
The difference between the definitional formula and computational formula
     for the variance and standard deviation
The advantages and disadvantages of using grouped data

3.5 Exercises

1. Write down a research project of your own interest, describing the variables
    you wish to collect. How would you summarize/describe your data? Would
    you compute the mean or the median for numerical variables? What other
    descriptive statistics would you provide?
2. Use the data from exercise 2 in chapter 2 (the heights in centimeters of a
    fictitious sample of 19 adult males in a research project). Compute the
    mean, median, mode, range, variance, and standard deviation with the
    data grouped and un-grouped.
3. You conduct a survey on ethnic membership in a small community, and
    find the following numbers. Which measure of central tendency should you
    use and why? Report it.

Ethnic group	f
Thais	30
Chinese	22
Laotian	11

4. You conduct a survey of a girls school's medical records in order to study
    the age at first menstruation (menarche) of a group of 10th graders
    who already had their menarche. The results are listed below. Group
    the data and compute the mean, median, mode, range, variance, and
    standard deviation. Let each year be a category in your frequency distri-
    bution.

9	16
10	16
14	11
11	10
11	13
12	13
14	14
13	10
13	11

16 12
15 12
14 13
11 14
12 15
14 13

4 | Probability and statistics

This chapter covers several topics on probability. Because probability is the foundation of statistical methods, this chapter provides the backbone for the rest of the book. Although important, this chapter is relatively short and to the point. My intention here is not to describe at length the laws of probability but to review how probability is related to scientific endeavor in general and to statistics in particular. R.A. Fisher (1959) eloquently explains the marriage between probability and scientific inquiry: mathematical reasoning is part of inductive reasoning in science. That is, with probability we can make statements about the likelihood of the occurrence of events. According to Fisher (1959: 110), scientific inferences involving uncertainty are accompanied by 'the rigorous specification of the nature and extent of the uncertainty by which they are qualified…'. In other words, in science we qualify statements about the occurrence of events with probability.

We will discuss the probability associated with discrete (qualitative and numerical discontinuous variables) and continuous data, and with sample means. However, before these topics are covered, we need to discuss the topic of sampling. Accordingly, this chapter is divided into five major sections:

1. random sampling and probability distributions,
2. the distribution of qualitative and discontinuous numerical variables,
3. the binomial distribution as used in biological anthropology,
4. the probability associated with continuous variables (including z scores and percentile ranks),
5. the probability distribution of sample means.

4.1 Random sampling and probability distributions

As just mentioned, this chapter deals with the probability associated with events of different kinds, the probability of obtaining a particular outcome in a particular population. But the first question we need to address is: how are we to sample from said population? Sampling should be random, that is, every individual should have an equal chance of being selected. But there is another aspect to random sampling: it can be done with or without replacement of subjects. An excellent example of the difference between these two forms of sampling is the 'random-play feature' present in compact disk players. Let's say that you can play five disks, and that each of them has five tracks. The probability of a specific track being selected is thus $1/(5)(5) = 1/25 = 0.04$. If the track is not selected in the first trial, and if the CD player samples *without* replacement, then the track that was already played will be taken out of the pool of possible tracks. Therefore, the probability of the specific song being chosen is now $1/24 = 0.042$. If it is not chosen the second time, then the track's probability is now $1/23 = 0.043$. As you can see, sampling without replacement alters the probability associated with events. In contrast, if the CD player samples with replacement, the probability associated with any one track would remain the same: 0.04 each time. It's perfectly possible, although certainly not likely, that the same track would be played again and again, since each track has the same probability of occurring each time.

When data are collected, a sample should be random and representative of the population. Thus, the researchers may wish not to replace subjects so that they get enough different variates to have an accurate representation of the population in the sample. For example, if an anthropologist is interested in interviewing female household heads to learn about the community's nutrition patterns, she would not want to interview the same female more than once (unless she wanted to check the reliability of her subjects' responses). Thus, the investigator would probably wish not to 'replace' her, that is, not to give her a chance of being selected again. As long as the population from which the sample is obtained is large enough, sampling without replacement does not substantially change the initial probabilities associated with specific events in the population. At the same time, sampling without replacement may allow the investigator to obtain a sample which better represents the population.

4.2 The probability distribution of qualitative and discontinuous numerical variables

There are many kinds of qualitative and discontinuous numerical variables in anthropological research. They will both be referred to as discrete data, since they can only take specific values (you can have blood type A, B, AB or O, but nothing in between; you can have 0, 1, 2, or more children, but not 0.75). Biological anthropologists frequently deal with hemoglobin and blood types, cultural anthropologists with different types of kinship labels, archaeologists with descriptive aspects of pottery assemblages. The purpose of computing the probability of observing a specific outcome (say a type of

Table 4.1. *Ceramic types from Depot Creek shell mound (86u56),* test *unit* A

Ceramic type	f	p
Sand-tempered	45	0.141 (45/318)
Grog-tempered	9	0.028 (9/318)
Check-stamped	184	0.58
Indent-stamped	28	0.088
Comp-stamped	8	0.025
Cord-Mark	1	0.003
Simple-stamped	43	0.135
	$n=318$	$\Sigma p=1$

Source: Data modified from White (1994)

hemoglobin) is to make inferences about the probability of sampling such outcomes from the population.

This topic is illustrated with an archaeological data set. In 1987 and 1988 White (1994) directed a number of excavations in various archaeological sites in the Apalachicola River Valley of Northwest Florida. The data in table 4.1 are the number of ceramic sherds by type found in test unit A at the Depot Creek shell mound (data modified from White, 1994). The table lists the number and percentages of ceramic type. Based on this information, we can predict the probability of finding different ceramic types in this site if we were to sample once more.

The probability p of sampling any of these outcomes is simply their frequency. Thus, the probability of sampling a check-stamped sherd is virtually 60%, whereas that of cord-stamped is virtually 0. Thus, the computation of expected outcome probabilities is useful when planning a follow up to a pilot study. If the researcher were to go back to the same site, and excavate in another area, and she found say, 60% cord, and 0.11% check-stamped sherds, she could conclude the following:

- a very unlikely event has occurred, but the true distribution of ceramic types in this site is still the one predicted by the table, or
- both test sites sampled a different population of ceramic types.

Practice problem 4.1

The following data show ceramic sherd types from Depot Creek shell mound (86u56) test unit B (White, 1994). What is the probability p of selecting all seven outcomes ? In your opinion, do test units A and B have similar enough frequencies to indicate that they were sampled from the same population?

Ceramic types from Depot Creek shell
mound (86u56) Test unit B

Ceramic type	f	p
Sand-tempered	40	0.18
Grog-tempered	15	0.07
Check-stamped	88	0.41
Indent-stamped	46	0.21
Comp-stamped	8	0.04
Cord-stamped	4	0.02
Simple-stamped	15	0.07
	$n=216$	$\Sigma p=1$

Source: Data modified from White (1994)

Although the two test units differ in their frequency of indent stamped pottery, they do appear to have a similar distribution. We shall pursue later the issue of how exactly to test for this difference. However, at this point the reader realizes that knowledge of a probability distribution (test unit A) allows researchers to make statements about the likelihood of finding specific outcomes in another sample of the same population.

The reader should recall that discontinuous numerical variables have only fixed values. Thus, when computing probability distributions of discontinuous numerical data, the procedure is basically the same as that applied to qualitative data. The following example illustrates this. Gray (1996) reports the number of children in a Turkana group who fall into eight age categories (measured in months). Table 4.2 shows part of the data presented by Gray in her table 1. The first category is not included here because it had a frequency of 0. Let us compute the probability associated with each outcome (age category in this sample).

These probabilities could be used to predict the likelihood of finding male children in each of these categories in a subsequent research project. The point to be stressed here is that discontinuous numerical variables, like quantitative ones, are discrete. Therefore, their frequency distributions are computed in the same manner.

4.3 The binomial distribution

A frequently used distribution for qualitative variables in biological anthropology is the binomial distribution. Because this distribution is most frequently used in biological anthropology, and not in the other sub-disciplines, it will be given limited treatment here.

Table 4.2. *Infant age distribution of
the Ngisonyoka of south Turkana; only
male infants shown*

Age category	f	p
1–3	14	0.27
4–6	8	0.15
7–9	11	0.21
10–12	4	0.08
13–15	10	0.19
16–18	2	0.04
>18	3	0.06
	$n=52$	$\Sigma p = 521.00$

Source: Data modified from Gray (1996)

As you probably know, human populations are polymorphic for many genetic systems. That is, human populations have more than one allele for a specific locus. A well-studied genetic system is that for hemoglobin (Hb). There are many human populations which have other alleles besides 'normal' hemoglobin, known as Hb A. The most widely known abnormal hemoglobin is Hb S (sickle hemoglobin), although there are other hemoglobins such as Hb C and Hb E.

Let us imagine that a population has the following allelic frequencies for hemoglobin: Hb A = 0.9, Hb S = 0.1 (thus, 0.9 + 0.1 = 1.0). When we talk about allelic frequencies we overlook that genes are carried by individuals, and visualize the population as a pool of genes, 90% of which consists of the A allele, and 10% of which consists of the S allele. Now we can ask: given these frequencies of alleles, what is the probability of sampling from this gene pool the following outcomes (genotypes or individuals): Hb AA (a homozygote A person), Hb AS (a heterozygote or carrier of Hb S), and Hb SS (a homozygote SS, or sickle cell anemic person)? We can visualize these probabilities as follows. An individual Hb AA has two A alleles, whose frequency is 0.9. Therefore, the probability of sampling such an individual is $(0.9)(0.9) = (0.9)^2 = 0.81$. The same can be said of the probability of sampling a sickle cell anemic patient. The probability of sampling an individual with two Hb S alleles is $(0.1)(0.1) = (0.1)^2 = 0.01$. Finally, the probability of sampling a heterozygous individual involves the sampling of an A and an S allele. But this can be done in either of two forms: A first and S second, or S first and A second. Thus, the probability of sampling a heterozygote is $(0.9)(0.1) + (0.1)(0.9)$, or $2(0.9)(0.1) = 0.18$. Therefore, the probability of sampling the three genotypes is Hb AA = 0.81, Hb AS = 0.18, and Hb SS = 0.01. Note that the probabilities add up to 1. The reader has probably noticed that what we have done is simply the expansion of the following binomial:

Formula 4.1 Computation of probabilities associated with genotypic frequencies

$$(p+q)^2 = p^2 + 2pq + q^2$$

Similarly, if we work in a population which has three alleles (say, Hb A, S and C), we can compute the expected probabilities of all possible genotypes by expanding $(p+q+r)^3 = p^3 + 3p^2q + 3pq^2 + q^3$. For higher binomial expansions, researchers can rely on Pascal's triangle, which provides the coefficients associated with higher level expansions (see Sokal and Rohlf, 1981).

Why do biological anthropologists use the binomial expansion? In studies of gene frequencies, the expected frequencies generated by the binomial expansions are those we would expect to find if the population is in Hardy–Weinberg equilibrium, in other words, if the population is not evolving and mates randomly. If we find a statistically significant difference between the observed and expected frequencies, we can conclude that the population is either experiencing evolutionary change or is not mating at random.

Practice problem 4.2

In an attempt to study the genetic structure of the present-day Mexican population, Lisker, Ramirez and Babinsky (1996) compute admixture estimates relying on gene frequencies thought to be representative of the parental populations. They present a table of gene frequencies of these ancestral populations, namely American Indian, African and European. For the American Indian population, they report the following gene frequencies for the MN system: M = 0.698, and N = 0.302. Compute the expected genotype frequencies.

If p is the frequency of the M allele, and q is the frequency of the N allele then

$$(p+q)^2 = p^2 + 2pq + q^2 = (0.698)^2 + 2(0.698)(0.302) + (0.302)^2 = 0.49 + 0.42 + 0.09 = 1$$

Notice how these expected frequencies could be used. If a biological anthropologist thought that these frequencies reflected the true frequencies of Mesoamerican populations before contact with Europeans and Africans (an assumption for this investigation), and if he wanted to test if an American Indian group had no admixture, then he would compare these expected frequencies with those observed in the population. If the researcher finds that the observed frequencies do not differ significantly from those expected, then he can conclude that this particular group does not differ from the ancestral population in its MN frequencies. If the observed and expected frequencies differ, the researcher would conclude that the population has experienced evolutionary change making it different from the ancestral population. Such evolutionary change could have resulted from gene flow, genetic drift, etc. In reality, this kind of study relies on many, not just one, genetic systems.

4.4 The probability distribution of continuous variables

The process of computation of expected frequencies of particular outcomes can also be applied to a sample of continuous quantitative data. However, continuous numerical data present a complication not encountered in either qualitative or discontinuous numerical variables: continuous numerical data can take an infinite number of possible values. Indeed, the number of categories recorded in a continuous data set is ultimately dependent on the accuracy of the measuring tool. Let us illustrate this complication with an example. Let us say that an anthropologist is collecting heights (in centimetres) of females attending a maternity clinic in an under-privileged neighborhood. The researcher's measuring tool only allows her to measure up to two decimal places. Below are her data (notice that this table does not have all the columns discussed in chapter 2 because the purpose here is not a full description of the data set).

Practical limits	Implied limits	Midpoint	f	$p(f/n)$
155.95–156.94	155.945–156.945	156.445	4	0.11
156.95–157.94	156.945–157.945	157.445	6	0.16
157.95–158.94	157.945–158.945	158.445	10	0.26
158.95–159.94	158.945–159.945	159.445	12	0.31
159.95–160.94	159.945–160.945	160.445	6	0.16
			$n=38$	$\Sigma p=1$

The question now is not what the probability is of finding a female whose height is (say) 156.445 but, rather, what is the probability of finding a female whose height lies between 155.95 and 156.94 (although we imply that the interval actually extends from 155.945 to 156.945). In our sample, the probability associated with this outcome is 0.11. We could use these data in the same way we used the probability distribution in the previous section, namely, as a base for predicting outcomes if we were to sample again. Thus, if the researcher samples heights of females of the same age group, but at a wealthy clinic, she has her previous study available for comparison. If she finds that instead of somewhere around 11% of the sample between 155.95 and 156.94, only 1% of her new sample falls in that category, she may be persuaded to think that the frequency distribution of heights is not the same in the populations from which these two samples were obtained (we still have not learned how to test for significant differences). The point that should be emphasized at this juncture is that it is possible to compute the frequency associated with outcomes in a continuous numerical frequency distribution. However, an outcome here is defined by two limits. This interval can be diminished in size, if the researcher wanted to know the probability associated with a more specific outcome. For example, the interval could be

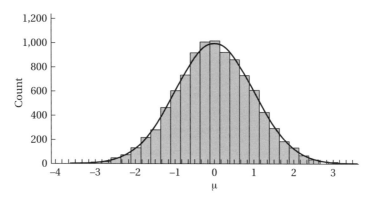

Figure 4.1. Distribution of a normal population, where $N=10,000$, mean $=0$ and standard deviation $=1$.

Figure 4.2. The normal curve illustrated with a sample of humans.

shortened to 156.95–157.44, with a midpoint of 157.195, and so on. However, this could become a tedious and difficult task, in which we fortunately do not have to engage (except for pedagogical purposes). The reader is probably familiar with the normal curve, whose bell shape is a frequent feature in statistics textbooks. The curve can be described mathematically by a function which requires more mathematical background than is assumed for this book, so the function is not presented here. Figure 4.1 was obtained by generating (with SAS) a normally distributed population of size $N=10,000$ whose μ is 0 and whose σ is 1. The histogram (of the population) closely follows the normal curve (generated by a mathematical function).

Statisticians and biologists have repeatedly shown that a large number of continuous numerical variables are normally distributed. Height is an excellent example of a normally distributed variable, as can be seen by figure 4.2. The picture shows a group of 175 humans, distributed according to their size. The frequency of subjects is highest in the middle range of the distribution, and lowest in the two tails.

If we work with a normally distributed variable, we can expect the following from the data (see figure 4.1).

Probability and statistics

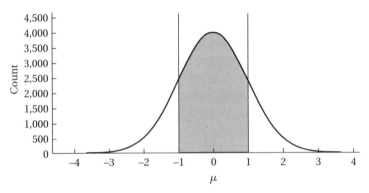

Figure 4.3. Distribution of a normal population, where $N = 10,000$. Area between ± 1 standard deviation units is 68.26% of the population.

1. *Symmetry.* The curve is symmetric. That is, 50% of all observations are above, and 50% below the mean μ.

2. *Frequency of outcomes.* The line delimiting the 'bell' actually tells us how frequent a particular outcome is. In other words, the higher the line associated with an outcome, the more likely the outcome is to occur. It is obvious that the outcome with the highest frequency is the mean (μ), and that the lowest occurs at the two tails, which approach asymptotically the horizontal axis (they never touch the axis). This is clearly exemplified by figure 4.2: most individuals are close to the average height; few are extremely short or extremely tall. But researchers can not be sure that they have measured the absolutely tallest or shortest person in any population. Thus, when we say that a variable is normally distributed, we say that most of the data take values close to the mean, with very few towards the tails. The probability of events approaches 0 the farther away from the mean the events are.

3. *Percentage of items under the curve.* Just as we know that 50% of the distribution is found to the right, and 50% to the left of the mean, we also know how items in each of the two halves are distributed. That is, we know that 34.13% of all items is found between the mean and the first standard deviation to its left, and 34.13% between the mean and the first standard deviation to its right. Thus, 68.26% (34.13 + 34.13) of all the population is found between the first negative and the first positive standard deviation (see figure 4.3).

We also know that 13.59% of outcomes is found between the first and second standard deviation (say, between −1 and −2). If this 13.59% is added to the 34.13% found between the mean and the first standard deviation, then 45.72% (13.59 + 34.13 = 45.72) of all the population is on one side between the mean and the second standard deviation. Since the distribution is symmetrical, then 95.44% (47.72 + 47.72) of all outcomes is found between the mean and the two standard deviations on both sides (see figure 4.4).

Finally, we know that 2.15% of all outcomes are found between the second and third standard deviations, which when added to the 47.72% of outcomes

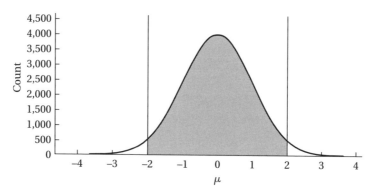

Figure 4.4. Distribution of a normal population, where $N=10,000$. Area between ± 2 standard deviation units is 95.44% of the population.

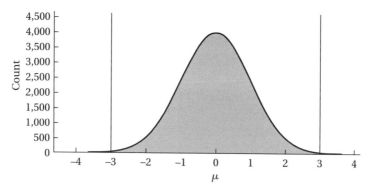

Figure 4.5. Distribution of a normal population, where $N=10,000$. Area between ± 3 standard deviation units is 99.74% of the population.

between the mean and the second standard deviation adds up to 49.87% on one side. Since the proportion is the same on both sides of the mean then $(2)(49.87) = 99.74\%$ of all outcomes is found between the mean and the third standard deviation on both sides (see figure 4.5).

The reader can see now why the curve never actually touches the horizontal axis: 0.26 % of all outcomes $(100 - 99.72)$ are found to the right and left of both third standard deviations. We can summarize this information as follows:

$\mu \pm \sigma$ contains 68.26% of all variates

$\mu \pm 2\sigma$ contains 95.44% of all variates

$\mu \pm 3\sigma$ contains 99.74% of all variates

4. *The probability associated with specific events in a continuous distribution:* recall how we estimated the probability of an event occurring in a qualitative data set. We determined that the frequency of the simple-stamped

ceramic type in test unit A in Depot Creek shell mound was $43/318 = 0.135$. We used this frequency to predict the probability of finding this ceramic type, if we were to sample from the same population. The probability of finding events in a normal distribution rests on the same principle and can be intuitively grasped by looking at figure 4.1. The reader can determine the frequency of an outcome by locating it in the horizontal line, then drawing a straight line up to the curve, and finding out the frequency pinpointed by the curve. But this is too cumbersome a method to find probabilities. Instead, we do the following: we transform a continuous data distribution into a standardized normal distribution whose $\mu = 0$ and whose $\sigma = 1$; the probability associated with specific outcomes has already been computed for such a distribution, and is listed in the unit normal table (table 1 in Appendix C). The table can readily answer the question we posed a few pages ago: what is the probability associated with sampling an individual whose height lies somewhere between 155.95 and 156.94, and who was sampled from a population whose μ and σ we know? Thus, we are going to first compute μ and σ, then transform the outcome whose probability of occurrence we wish to know into a 'z score', or a 'standardized normal deviate', and use table 1 to estimate the probability associated with this outcome. This entire process is described in the next section.

4.4.1 z scores

The use of **z scores** is to facilitate the process of calculating the probability associated with outcomes in a continuous distribution. A z score is a specific outcome which has been subtracted from its population mean, and divided by its population standard deviation. Thus, the outcome is now said to be standardized. The reader recalls that many variables *are* normally distributed. However, each variable has its own μ and σ, which makes the computation of such probabilities cumbersome each time they are needed. The solution is thus to use a standardized normal distribution whose mean is 0 and whose standard deviation is 1. All we have to do is to transform the outcome of our interest into a z score, and determine its probability of occurrence with the normal unit table. If we transform an entire data set into z scores, the data would now have a mean of 0 and a standard deviation of 1.

The computation of a z score is quite simple, but requires knowledge of the population parameters. The formula for a z score is:

Formula 4.2 Formula for the computation of a z score

$$z = \frac{Y - \mu}{\sigma}$$

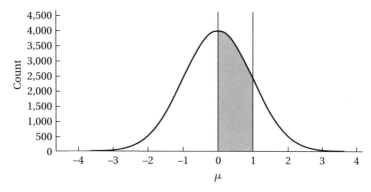

Figure 4.6. Area between the mean and a z score whose value is 1. This proportion is given by column B of table 1 in Appendix C.

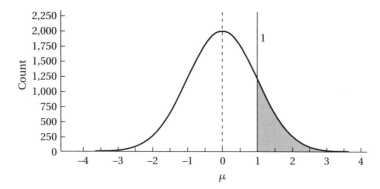

Figure 4.7. Area of the curve beyond a z score whose value is 1. This area is given by column C of table 1 in Appendix C.

Here, Y is the outcome whose probability of occurrence we wish to ascertain. For example, if our population of females has a mean height of $\mu = 160\,\text{cm}$ and a standard deviation of $\sigma = 7$, the z score for a female whose height is 156 cm is

$$z = \frac{156 - 160}{7} = -0.57$$

Now we can learn how to use table 1. It has three columns: column A lists the z scores, column B the area of the curve between the mean and the z score, and column C the area of the curve beyond the z score. These areas are illustrated in figures 4.6 and 4.7.

Before we ask how likely we are to find a female whose height is 156, we will ask some preliminary questions. The first thing we need to do is to locate the z score associated with that outcome, namely, $z = -0.57$. A quick view of column A shows that there are no negative scores. This is because a normal distribution is symmetrical, so we can answer all our questions by looking at the positive side of the distribution. Thus, we find the z score 0.57. With columns B and C we can answer two simple questions. What is the proportion of the distribution between the mean and a subject whose height is 156 (and whose z score is −0.57)? Column B answers that question: 0.2157%. What is the probability of finding a female who is as short, or shorter, than our female? Column C answers that question: 0.2843%. When working with z scores, it is useful to draw the normal distribution, marking the area of inquiry, as done in the following practice problems.

Practice problems 4.3

Given that our population of females has a mean stature of $\mu = 160$ and a standard deviation of $\sigma = 7$, answer the following questions:

1. What is the probability of finding a female whose height falls between 154 and 167? Note that these two observations are on opposite sides of the mean (which is 160). Thus, we need to find the area between the mean and the z scores, which is provided by column B, and sum them. First we compute the z scores:

$$z = \frac{154 - 160}{7} = -0.86, \quad \text{and} \quad z = \frac{167 - 160}{7} = 1.0.$$

Then we find the values of column B for both z scores: For z = 0.86 the area is 0.3051, and for z = 1 the area is 0.3413. We sum them for the answer: the probability of finding a female whose height falls somewhere between 154 and 167 is 0.3051 + 0.3413 = 0.6464. This is illustrated in the following graph:

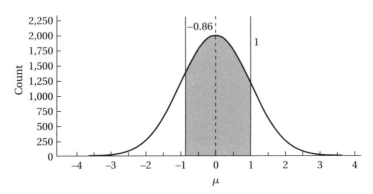

Figure 4.8. Area of the curve between −0.86 and 1.

2. What is the probability of sampling a female whose height is between 170 and 180 cm? Both outcomes are on one side of the distribution, and we want to find the probability delimited by both z scores. To find it, we need to subtract the smaller column B value from the larger one. We proceed as follows:

z scores   Column B   Answer

$$z = \frac{170 - 160}{7} = 1.43 \quad 0.4236$$

$$z = \frac{180 - 160}{7} = 2.86 \quad 0.4979 \quad 0.4979 - 0.4236 = 0.0743$$

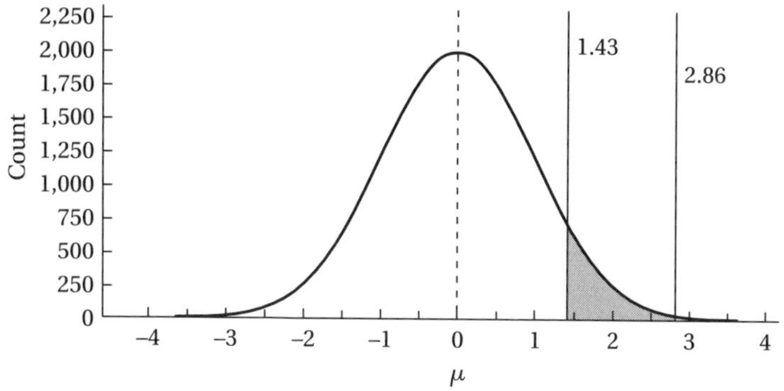

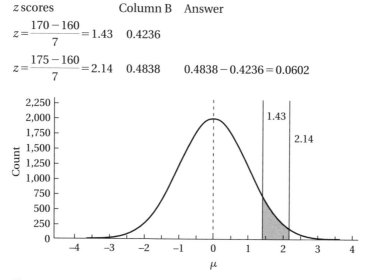

Figure 4.9. Area of the curve between two z scores whose values are 1.43 and 2.86.

3. Let us now narrow our inquiry: what is the probability of sampling a female whose height is between 170 and 175?

z scores   Column B   Answer

$$z = \frac{170 - 160}{7} = 1.43 \quad 0.4236$$

$$z = \frac{175 - 160}{7} = 2.14 \quad 0.4838 \quad 0.4838 - 0.4236 = 0.0602$$

Figure 4.10. Area of the curve between two z scores whose values are 1.43 and 2.14.

4. What is the probability of sampling a female whose height is between 170 and 171?

z scores   Column B   Answer

$$z = \frac{170 - 160}{7} = 1.43 \quad 0.4236$$

$$z = \frac{171 - 160}{7} = 1.57 \quad 0.4418 \qquad 0.4418 - 0.4236 = 0.0182$$

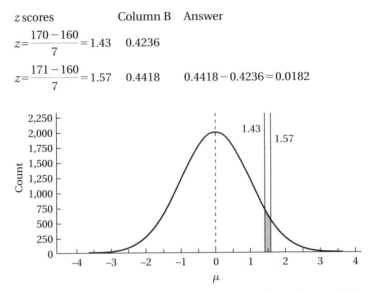

Figure 4.11. Area of the curve between two z scores whose values are 1.43 and 1.57.

4.4.2 Percentile ranks and percentiles

We have all had to deal with **percentiles**. Graduate students take the GRE (Graduate Record Examination), and would much rather be in the 90th than in the 50th percentile. The proud parents of babies eagerly await the offspring's monthly visit to find out what height and weight percentile the baby is in. Percentiles are indeed a popular use of z scores. The **percentile rank** is the percentage of individuals in the distribution with scores at or below that particular score. The specific score associated with a percentile rank is called a percentile. If we say that a test score was in the 50th percentile, then it fell at the mean of the distribution. If it is in the 60th percentile, 60% of the distribution is below it.

To find out what percentile an observation falls in, it is transformed into a z score. Let us say that in a population of test scores whose $\mu = 110$ and $\sigma = 12$, a student gets a score of 140. What percentile is associated with this datum? We compute a z score as always:

$$z = \frac{140 - 110}{12} = 2.5.$$

The test score is higher than the mean, so at least we know that it is higher than the 50th percentile. Then we look for the value of column B associated with $z = 2.5$, which is 0.4938. If we add that to the 50% to the left of the mean, we find out that the score in the $0.5 + 0.4938 = 99$th percentile.

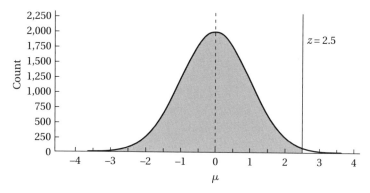

Figure 4.12. Percentile rank of a score of 140 in a population whose $\mu = 110$ and standard deviation $= 12$.

What if the test score was 90? The z score would be:

$$z = \frac{90 - 110}{12} = -1.67.$$

We need to use column C here, because we are looking for the proportion of the distribution with values under this grade. For $z = -1.67$, column C $= 0.0475$. Sadly, the grade is in the 4th percentile.

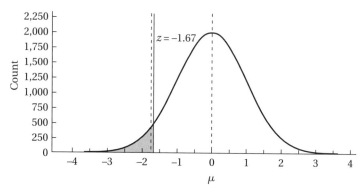

Figure 4.13. Percentile rank of a score of 90 in a population whose $\mu = 110$ and standard deviation $= 12$.

4.5 The probability distribution of sample means

This section deals with an application of probability in statistics that is more relevant to actual research. Let us say that a cultural anthropologist is investigating the age at which females marry in a population. The anthropologist ascertains that the mean sample age at marriage is 20 years. But, how representative is this $\overline{Y}$ from the population's μ? This section will cover the probability of obtaining a sample of size n with a specific $\overline{Y}$, *assuming* knowledge of the population's μ and σ. Obviously, this is an unrealistic situation, since we rarely have such population knowledge. However, probability deals with hypothetical situations and, for us to understand statistical reasoning, we need to understand the workings of probability in such hypothetical scenarios. In the next chapter we will make the transition to the 'real world'.

Let us continue our discussion with an example of a population of only $N = 3$ subjects whose values are: $Y_1 = 10$, $Y_2 = 20$ and $Y_3 = 30$. In other words, we have total knowledge of the population. Let us now proceed to sample with replacement all possible samples of size $n = 2$, and compute for each sample its $\overline{Y}$, as well. Since we are sampling with replacement, the sample could consist of the same score twice. Starting with Y_1, the samples could include Y_1 and Y_1, Y_1 and Y_2, and Y_1 and Y_3. The following table shows the list of all possible samples and their sample means. Also shown is a frequency distribution of the sample means and a bar graph of their values. It should be apparent that, if all possible samples of size n are sampled from a population, the sample means are normally distributed. Notice that our population was most certainly not normal: it had a flat 'curve', with three subjects each of which was found only once (instead of a normal bell curve, with the highest frequency found at the mean). However, the distribution of the sample means is normal.

Sample number	First observation	Second observation	$\overline{Y}$
1	10	10	10
2	10	20	15
3	10	30	20
4	20	10	15
5	20	20	20
6	20	30	25
7	30	10	20
8	30	20	25
9	30	30	30

The frequency distribution of the sample means is:

Sample means	f	p
10	1	$1/9 = 0.111$
15	2	$2/9 = 0.222$
20	3	$3/9 = 0.333$
25	2	$2/9 = 0.222$
30	1	$1/0 = 0.111$
$\mu = 20$	$\Sigma f = 9$	$0.999 \approx 1$

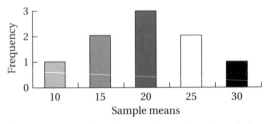

Figure 4.14. The distribution of sample means of a known population.

We can now go a step further. Given that we have a distribution of sample means, we can compute the probability associated with particular outcomes (sample means) as done previously in this chapter. Such probabilities are also shown above. Thus, what is the probability of obtaining a sample of $n = 2$ with a mean of $\overline{Y} = 10$? Since only one sample had this mean, the probability is easily computed by dividing $1/9 = 0.11$, where $\Sigma f = 9$ is the total number of samples. What is the probability of obtaining a sample of size $n = 2$ with a mean of $\overline{Y} = 20$? The probability of such an outcome is $p = 0.333$. Therefore, when we have total knowledge of a population, and can obtain all possible samples of size n, we can quantify the probability associated with obtaining a specific outcome or $\overline{Y}$. Imagine what a cumbersome task that would be if instead of having a population size $N = 3$, and instead of obtaining all possible samples of size $n = 2$, our population were of size $N = 1,000$ and we wanted to know the probability distribution of samples with size $n = 30$. Once again, we don't engage in such onerous task but instead rely on z scores. Instead of computing a probability distribution and the probability associated with specific outcomes as we did above, we can transform our sample mean into a z score, and use table 1 to find the probability associated with such an outcome.

As the reader knows, to compute a z score we need to divide the difference between $\overline{Y}$ and μ by some kind of standard deviation. However, what we need in this case is the standard deviation, not of a sample of individuals, but a standard deviation of a sample of sample means. The value of the standard deviation of the sample means depends on two items: the population's standard deviation σ (which is the standard deviation of the population's subjects), and the size of the sample. It has been shown that, if all possible samples of size n are taken from a known population, their scatter will be greater as their sample size diminishes. Conversely, as n increases, the sample means will scatter closer to the population's μ. *Assuming knowledge of the population*, the formula for computing the standard deviation of means is:

Formula 4.3 The standard deviation/error of the means

$$\sigma_{\overline{Y}} = \frac{\sigma}{\sqrt{n}}$$

If $\sigma = 8.16$ and $n = 2$ then $\sigma_{\overline{Y}} = \dfrac{8.16}{\sqrt{2}} = 5.77$. If, however, $n = 5$ then $\sigma_{\overline{Y}} = \dfrac{8.16}{\sqrt{5}}$ $= 3.64$, and if $n = 10$ then $\sigma_{\overline{Y}} = \dfrac{8.16}{\sqrt{10}} = 2.58$. Clearly, if sample sizes are increased, the distribution of the means will be less scattered, more clustered toward the population's mean.

Please note that the standard deviation of the means is frequently referred to as the standard error of the means, perhaps a better descriptor of the parameter. Because the larger the value of $\sigma_{\overline{Y}}$ the more scattered the sample means will be from the population mean, the more error there will be in the estimation of the population parameter. Conversely, if the standard error of the means is small, the sample means will be closer to the population parameter, and the error in the estimation of the population parameter will be diminished. In this book we will use interchangeably the terms standard deviation of the means and the standard error of the means.

We are now in a position to answer the question we asked at the beginning of this section: assuming knowledge of the population, what is the probability of obtaining a sample with a specified n and $\overline{Y}$? We will answer that question by computing a z score which uses $\sigma_{\overline{Y}}$ instead of σ. Thus, our z score formula is slightly changed to:

....

Formula 4.4 Formula for the z score of a sample mean

$$z = \frac{\overline{Y} - \mu}{\dfrac{\sigma}{\sqrt{n}}} \quad \text{or } z = \frac{\overline{Y} - \mu}{\sigma_{\overline{Y}}}$$

....

Practice problem 4.4

Shackley (1995) reports elemental data analysis of several obsidian sources from the greater American Southwest. He presents (in parts per million) the concentration for several elements, among them titanium (Ti). For this exercise, let us assume that the mean and standard deviations reported by Shackley are the parameters μ and σ. For Cow Canyon Arizona, Shackley reports $\mu = 1{,}067.20$ and $\sigma = 226.80$, and for the Mule Mountain group $\mu = 720.45$ and $\sigma = 105.29$. Both sites are in the Eastern Arizona/Western New Mexico area, which is where our fictitious excavation takes place.

The characteristics of our sample are the following: $n = 15$, $\overline{Y} = 850$. Compute z scores for our sample in order to find out the probability of obtaining a mean at least as deviant as ours in both populations (thus, we need to use column C). What do the results suggest about the origin of our sample?

We compute the z score using the Cow Canyon data first:

$\sigma_{\overline{Y}}$	z score	Probability from column C
$\sigma_{\overline{Y}} = \dfrac{226.80}{\sqrt{15}} = 58.56$	$z = \dfrac{850 - 1{,}067.20}{58.56} = -3.70$	0.0001

According to the information provided by column C, the probability of obtaining a sample mean of 850 or less in this population would be 0.0001, a low probability as demonstrated in figure 4.15.

We then compute the z score with the Mule Mountain group data:

$\sigma_{\overline{Y}}$	z score	Probability from column C
$\sigma_{\overline{Y}} = \dfrac{105.29}{\sqrt{15}} = 27.18$	$z = \dfrac{850 - 720.45}{27.18} = 4.76$	≈ 0

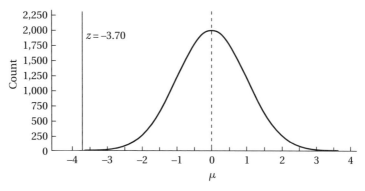

Figure 4.15. Probability associated with obtaining our mean if the sample came from the Cow Canyon population.

A z score whose value is greater than 4 is so far removed from the mean that the probability of finding it, or anything more extreme, approaches 0.

This example is noteworthy because it demonstrates that the difference between an outcome and the mean of a distribution alone can not answer our question about the probability of sampling such an outcome. Obviously, the absolute difference (without regard to the sign) between our sample mean and the mean of the Mule Mountain group is smaller $(850 - 720.45 = 129.55)$ than the difference between the sample mean and the Cow Canyon mean $(850 - 1067.20 = -217.2$, whose absolute value is 217.2). However, the latter population has a larger standard deviation which yields a larger standard error of the mean. This means that the distribution of sample means of the Cow Canyon population is more scattered, allowing greater variation than that of the other population.

In this and previous examples, we dealt with hypothetical situations, in which we assumed knowledge of the population parameters. In the next chapter we shall deal with 'real life' situations, in which we do not have knowledge of the population's parameters. Perhaps the most important concept of the entire chapter is that probability provides the foundation upon which we can make scientific statements. A person who says that there is 'scientific proof for the occurrence of an event…' is actually saying that 'the probability associated with the event is close to 1'. We can not say that an obsidian sample did not come from population A and did come from population B. We can only say that the probability associated with obtaining such a sample from population A approaches 0, but it is greater than 0 if the sample came from population B. Thus, z scores can be used to predict the likelihood of specific outcomes.

4.6 A research example of z scores

Leatherman (1994) conducted a study to investigate how agrarian reform may have affected the health status of rural populations in highland

Peru. He measured nutritional status in three sites (semi-urban town, agrarian community and herding cooperative) by recording anthropometric data. Leatherman wished to determine if the children in the three communities were growth-stunted and, if so, which community is more stunted. In this case z scores are of much use: the children from the communities can be compared with one distribution, namely, the NCHS reference population. Thus, each child's datum (height-for-age) can be converted into a z score using the National Center for Health Statistics's (NCHS) μ and σ. Children whose z scores are less than or equal to -3 are considered to be suffering chronic undernutrition. The author finds that 18.5% of the town, 31.7 % of the community, and 15.9% of the cooperative children suffer from chronic malnutrition. According to Leatherman (1994) these data show that, whereas the herding community benefited the most from the reform, and the farming commune the least, the people from the town fell in between. This is an excellent example of how z scores can be used to place a datum (say, a child's height-for-age) in the distribution of a population in order to determine where in this distribution the observations falls.

4.7 Key concepts

Random sampling
Probability distribution of qualitative or discrete discontinuous variables
Probability distribution of continuous variables
Binomial distribution
Distribution of a normal population
Asymptotic approach to the horizontal axis
z score
Percentiles and percentile ranks

4.8 Exercises

1. Describe ways to ensure procurement of a random sample. If you choose not to use replacement, justify this in the context of a research situation.
2. Compute the expected genotype frequencies for the following hemoglobin data on a fictitious sample in a tropical population:
   HbA   HbS
   0.92   0.08
3. A social marketing researcher has sampled a group of older women from a diverse community, to determine if they are having yearly mammograms. The data are shown below. You have been hired to do a follow-up study in the same community, in which you will sample a total of 80 females. Given the ethnicity of the females who were sampled already, construct a

probability distribution, and compute the probability associated with each ethnic group. Given these probabilities, how many females (out of 80) would you expect to find in your new sample, who belong to each of these ethnic groups?

Participant 1 Hispanic
Participant 2 Hmong
Participant 3 Hispanic
Participant 4 African American
Participant 5 African American
Participant 6 Japanese
Participant 7 Hispanic
Participant 8 African American
Participant 9 Hispanic
Participant 10 African American
Participant 11 Hispanic
Participant 12 Greek
Participant 13 African American
Participant 14 Hispanic
Participant 15 Chinese
Participant 16 African American
Participant 17 African American
Participant 18 Hispanic
Participant 19 African American
Participant 20 Samoan
Participant 21 Hispanic

4. It is known that, in a particular archaeological population, the mean number of linear enamel hypoplasias (defects in enamel development due to a metabolic insult to the organism) per individual is $\mu = 5$ with $\sigma = 0.8$. What is the probability of excavating:
   (a) an individual with one or fewer lesions,
   (b) an individual with between three and six lesions,
   (c) an individual with between two and four lesions,
   (d) an individual with nine or more lesions.

5. It is known that the elderly women of a certain rural American population have a mean weight of $\mu = 150$ pounds, with a standard deviation of $\sigma = 10$. What is the probability of finding in such population:
   (a) a woman between 125 and 135 pounds?
   (b) a woman between 140 and 160 pounds?
   (c) a sample of size 10 with $\overline{Y} = 170$? What would your answer be if $\sigma = 7$?
   (d) a sample of size 20 with a $\overline{Y} = 170$? What would your answer be if $\mu = 7$?

6. One series of archaeological tapestry was woven in a style peculiar to a specific time period. It has a mean length of $\mu = 0.9$ meters, with $\sigma = 0.4$

meters. A second population of tapestry known to have been woven during the same time, at a nearby area, has a mean length $= 1.5$ meters and $\sigma = 0.5$ meters. You have unearthed a sample of tapestry from the same time period. There are $n = 11$ tapestries in your sample, with $\overline{Y} = 1.4$ meters. Compute a z score for the sample twice, using both populations' data. In your opinion, to which population does the sample belong?

5 Hypothesis testing

In chapter 1 we discussed the use of statistical analysis in the scientific endeavor. Not only do statistics allow us to summarize and understand our data, but they permit us to test hypotheses about the population from which data are collected. The marriage between statistical analysis and scientific-hypothesis testing is elegantly expressed by G.M. Jenkins (1979): 'The pre-occupation of some statisticians with mathematical problems of dubious relevance to real-world problems has been based on the mistaken notion that statistics is a branch of mathematics – in contrast to the more sensible notion that it is *part of the mainstream of the methodology of science*' (italics in original). Fisher (1959) mentions that the people who developed (now classic) statistical tests did so in the midst of a research problem. The statistics were developed to answer a scientific need. This chapter introduces the reader to principles of hypothesis testing, and how statistics are used for this scientific purpose. Examples of such tests are included.

5.1 The principles of hypothesis testing

Earlier in the book it was mentioned that a scientific hypothesis is one which can be tested. This testing proceeds (basically) in the following manner.
1. State the null (H0) and alternative (H1) hypotheses.
2. Establish the level of statistical significance.
3. Collect the sample.
4. Compare the sample with the null hypothesis, and reach a conclusion about which hypothesis to accept.
Each of these points is discussed below.

1. State the null (H0) and alternative (H1) hypotheses. The first step in hypothesis testing is for the researcher to propose a null hypothesis about the population under study, specifically about the population's parameters. An example of such a null hypothesis could be: 'In this population, the mean age at marriage is $\mu = 20$ years.' Different books list different reasons about why H0 is called the null hypothesis. Some authors (Sokal and Rohlf, 1981) indicate that the H0 receives its name from the fact that it proposes that there is no difference between the true value of the population's parameter and that which is proposed in the H0. Other authors (Gravetter and Wallnau,

1992) mention that the H0 is called the null hypothesis because it states that the population parameter has not been affected by a treatment effect, in other words, that even if a population has been subjected to a treatment, said treatment has had a null effect on the population. Both reasons for H0's name are valid.

The alternative or scientific hypothesis (H1) proposes that the treatment has had an effect, and that the true parameter of the population from which our sample was obtained is different from that proposed by the null hypothesis. For example, a researcher may know that the mean age at marriage in a country is $\mu = 20$ years. However, he is interested in studying the age at marriage in a community which is ethnically distinct from the larger national culture. The anthropologist suspects that cultural practices such as age at marriage differ in this small community from those of the larger national culture. The researcher then proposes the following hypotheses: H0: $\mu = 20$ years. H1:$\mu \neq 20$ years. The alternative hypothesis here proposes that the treatment effect (ethnicity) has a significant effect on the population, making the small community's μ *statistically significantly* different from 20.

It is interesting that, even though H1 is the hypothesis generated by the research interest, H0 is the one being tested. If H0 is not rejected, it is accepted as the best explanation of the facts, but it has not been proven. If H0 is rejected, H1 is accepted as the best explanation of the facts, but it has not been proven either. Indeed, this is how scientific knowledge progresses: researchers accept the currently most likely explanation of the facts, while being totally open about the possibility that such explanation may be falsified in the future.

A very important topic in hypothesis testing and in science in general is discussed by S.J. Gould in his essay titled 'Cordelia's dilemma' (1993). The issue concerns the (very human) desire to 'prove our hypothesis' (although we never prove anything in science). This is a very understandable desire: if a researcher spends much time and effort researching a topic, thinking that (H1) X affects Y, only to find out that it does not, she will surely be disappointed. But what we all need to realize is that negative results (H0 is not rejected) are as important as positive ones (H0 is rejected). Both kinds of results inform us about the behavior of our statistical population. Indeed, perhaps by only concentrating on positive results we get a biased view of the actual nature of our population.

2. Establish the level of statistical significance. Before the researcher collects a sample to test his hypothesis, he must decide how different a sample mean $\overline{Y}$ must be from the population proposed by H0 for the H0 to be rejected. The investigator faces the following problem: if he is to sample repeatedly from a population, the mean of different samples will likely differ. Indeed, in the previous chapter we performed an 'experiment' in which we obtained all possible samples of size $n = 2$ from a known popula-

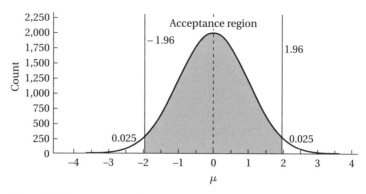

Figure 5.1. Acceptance and rejection regions of the normal curve.

tion of size $N=3$ and whose $\mu=20$. We *know* that every single of these samples *was* collected from the population. However, whereas some sample means were equal to the population mean, others were rather different. Therefore, the problem in hypothesis testing is to establish how deviant a sample mean must be from the μ proposed by the H0 for the researcher to decide to reject H0. The solution to this problem is the following: we compute a z score for the $\overline{Y}$ *using the population information proposed by the H0,* and see where in the population such a z score falls. It is a widely accepted statistical convention that if a z score falls within the 95% of a population's distribution around the μ, it is considered not to differ in a statistically significant manner from the population's mean. The remaining 5% of the distribution, equally divided between both tails of the distribution (2.5% on each side), is considered to be significantly different from the population's mean. Accordingly, 95% of the distribution is called the acceptance region (if a z score falls within it, the H0 is accepted), and the 5% of the distribution, equally divided between both tails, is called the rejection or critical region. The rejection area is referred to as alpha (α), and its value is established by the researcher to be (usually) 5% or 0.05. The acceptance region is obtained by subtraction: $1-\alpha$.

We will initially use table 1 in Appendix C to test hypotheses: what we need to find is a z score whose column C value is 0.025, since we are equally dividing 0.05, the level at which we will reject H0 between both tails. Such a z score is 1.96, whose column C value is 0.025. Therefore, if a z score is equal to or greater than 1.96, it will be considered to be significantly different from the population's mean.

The reader is reminded that we are discussing statistical as opposed to anthropological/biological/cultural significance. These two may or may not be the same. Indeed, this is related to the importance to the scientific endeavor of negative, non-statistically significant results. It is possible that, if a result is considered of non-statistical significance (the null hypothesis is not rejected), it may still give us great insight into our population. Bramblett

(1994) discusses that, in Primatology, behavioral/biological significance may be divorced from statistical significance. If one rare event marks the change of an individual's status in a group's hierarchy, we still need the many hours of observation in which this occurrence did not occur for us to understand how significant this rare event was to the group. The same can be said of paleoanthropological studies: even if we are interested in the brief, episodic, and significant periods of directional change in the fossil record, we still need to document and understand the long periods of no directional change (Gould, 1993).

3. Collect the sample. After the H0 and H1 have been proposed, and the researcher has chosen a particular level, the sample may be collected. As mentioned previously, a discussion of research design and data collection is outside the scope of this book.

4. Compare the sample with the null hypothesis, and reach a conclusion about which hypothesis to accept. At this stage, the researcher computes the sample mean's z score as done in the last chapter $(z = (\bar{Y} - \mu)/\sigma_{\bar{Y}})$. If the z score is greater than or equal to 1.96 then it will be considered to be significantly different from the μ proposed by the H0.

In our fictitious research project on age at marriage in a small ethnically distinct community we proposed the following hypotheses: H0: $\mu = 20$ years H1: $\mu \neq 20$ years. We know from census data that for the entire country $\mu = 20$ and $\sigma = 5$ years. What would our decision be if we collected a sample of $n = 30$ whose $\bar{Y} = 19$? What would our decision be if we collected a sample of $n = 30$ whose $\bar{Y} = 18$?

$\bar{Y}$	n	$\sigma_{\bar{Y}}$	z	Decision
19	30	$\dfrac{5}{\sqrt{30}} = 0.91$	$\dfrac{19 - 20}{0.91} = -1.1$	Accept H0
18	20	$\dfrac{5}{\sqrt{30}} = 0.91$	$\dfrac{18 - 20}{0.91} = -2.2$	Reject H0

If the sample mean had been 19, we would decide that the mean of the population from which we obtained our sample was indeed $\mu = 20$. A sample mean of 18 would be considered to be too different from $\mu = 20$ (its z score is greater than 1.96), and would be declared not to have been sampled from the population proposed by H0.

Practice problem 5.1

For this exercise, our population's parameters are $\mu = 20$ and $\sigma = 5.77$. The population size is $N = 3$ with $Y_1 = 10$, $Y_2 = 20$, and $Y_3 = 30$. Let us obtain five different samples of $n = 2$ whose respective means are 10, 15, 20, 25, and 30. Let us compute for each sample mean a z score to test the null hypothesis that the population from which these samples were obtained has a mean of 20. Thus: H0: $\mu = 20$; H1: $\mu \neq 20$; If the z score is equal to or greater than 1.96, we reject H0.

$\overline{Y}$	n	$\sigma_{\overline{Y}}$	z	Decision
10	2	$\dfrac{5.77}{\sqrt{2}}=4.1$	$\dfrac{10-20}{4.1}=-2.44$	Reject H0
15	2	$\dfrac{5.77}{\sqrt{2}}=4.1$	$\dfrac{15-20}{4.1}=-1.22$	Accept H0
20	2	$\dfrac{5.77}{\sqrt{2}}=4.1$	$\dfrac{20-20}{4.1}=0.0$	Accept H0
25	2	$\dfrac{5.77}{\sqrt{2}}=4.1$	$\dfrac{25-20}{4.1}=1.22$	Accept H0
30	2	$\dfrac{5.77}{\sqrt{2}}=4.1$	$\dfrac{30-20}{4.1}=2.44$	Reject H0

This example is of pedagogical use: it illustrates the relationship between hypothesis testing and probability theory. All other hypothesis tests presented in this book rest on the same principles: if the outcome of an experiment or survey is statistically unlikely to have been obtained from a population, then it is said not to have come from such population. Unlikely here is defined as falling in the rejection area. Clearly however, our decision to reject the null hypothesis in this practice was in error: we *know* that the samples with $\overline{Y}=10$ and 30 did come from the population. The next section discusses errors in hypothesis testing.

5.2 Errors and power in hypothesis testing

When testing a hypothesis, a researcher could make the following errors:

	Null hypothesis is true	Null hypothesis is false
Decision: reject H0	Type I error	Correct decision
Decision: accept H0	Correct decision	Type II error

5.2.1 Type I error (α)

A **type I error** was illustrated in the previous section: the null hypothesis was correct even with the samples whose $\overline{Y}=10$ and 30, yet we decided to reject H0 because these outcomes fell outside the acceptance region. How likely are we to commit this error? The answer is simply that the probability is equal to , in this case 0.05. Therefore, when we decide to reject a null hypothesis, we also implicitly state that we know that we could be making an error in our decision, and we actually state how likely we are to be making such an error: the probability is α. This means that if $\alpha=0.05$ then 1 out of 20 trials $(1/20=0.05)$ or samples obtained from the population could lead the researcher to erroneously reject H0. If the reader thinks this is too high a risk, then she could decide to decrease her rejection region to say, 0.01. This would mean that now 1 out of 100 samples could lead to a type I error. The reader could go even further and use an alpha level of 0.001, in which case only 1 out of 1,000 samples would result in a type I error. The z scores associated with these levels are:

Alpha	z score	Column C
0.05	1.96	0.025
0.01	2.58	$0.0049 \approx 0.005$
0.001	3.3	0.0005

However, the reader has probably grasped by now that this exercise, while decreasing α, has only increased the probability of committing a type II error (accepting H0 when in fact it is false). This is why an α level of 0.05 is considered a good middle ground.

5.2.2 Type II error (β)

The topic of what exactly the probability associated with **type II error** or β is frequently left out in statistics textbooks. This is because in most cases such probability cannot be computed since it requires the H1 to be stated as precisely as the H0 is. For example, in our age-at-marriage project, we would need to state H0:$\mu = 20$ and (e.g.) H1: $\mu = 25$. As the reader can see, this level of precision requires knowledge about two populations, knowledge which is usually unavailable. However, the probability of committing a type II error is illustrated here with another purely pedagogical example.

Let us assume that we have total knowledge of two populations, both of which have size $N = 3$. The first population's observations are 10, 20 and 30. The second population's observations are 15, 25, and 35. If we sample with replacement all possible samples of size $n = 2$ we obtain the following frequency distributions and histograms (figure 5.2) for the sample means of both populations:

Population 1:H0			Population 2: H1		
Sample means	f	p	Sample means	f	p
10	1	$1/9 = 0.111$	15	1	$1/9 = 0.111$
15	2	$2/9 = 0.222$	20	2	$2/9 = 0.222$
20	3	$3/9 = 0.333$	25	3	$3/9 = 0.333$
25	2	$2/9 = 0.222$	30	2	$2/9 = 0.222$
30	1	$1/0 = 0.111$	35	1	$1/0 = 0.111$
$\mu = 20$	$\Sigma f = 9$	$0.999 \approx 1$	$\mu = 25$	$\Sigma f = 9$	$0.999 \approx 1$

In this example we can compute the probability of a type II error, which is simply the proportion of H1's population which overlaps with the acceptance region of H0's distribution. The reader may recall that the acceptance region of H0's distribution included only samples whose means are 15, 20 and 25, all of which overlap with H1's distribution. If the probabilities of those sample means are added, we obtain the probability of computing a type II error. Thus, $\beta = 0.222 + 0.333 + 0.222 = 0.777$.

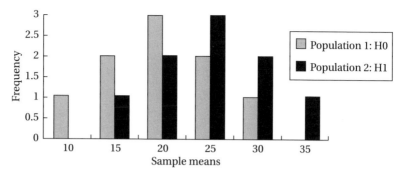

Figure 5.2. The sampling distribution of two populations (H0 and H1).

5.2.3 Power of statistical tests (1 − β)

The previous section brought us to the topic of **power** in a statistical test, which is the ability of a test to reject H0 when H0 is false. Since a type II error (β) is the acceptance of H0 when in fact it should be rejected, power is defined as $1 - \beta$. Therefore, in the example presented in the previous section, the power of our test was $1 - 0.777 = 0.223$. Statistical power, just like the probability of a type II error, depends on how different the means of the populations proposed by H0 and H1 are. If H0: $\mu = 20$ and H1: $\mu = 20$ then the overlap of H1's distribution with the acceptance region of H0 is total. In this case, β would be equal to the entire acceptance region, which (if $\alpha = 0.05$) would be $1 - 0.05 = 0.95$. The more different the two parametric means are, the higher the power of the test.

The computation of β and $1 - \beta$ is slightly more complicated if we are working with the distribution of continuous data. In this case, we need to rely on the normal distribution table to establish the proportion of H0's population acceptance region which overlaps with H1's distribution. We proceed as follows.

1. Establish the 95% cut-off points of H0's distribution.
2. Compute a z score by subtracting from such cut-off point the value of H1's μ, and dividing by the standard error of the means.
3. Column C of table 1 tells us the proportion of the distribution beyond the z score. This is the proportion of H0's acceptance region which overlaps with H1's distribution. This procedure is illustrated below.

Figure 5.3 shows the distribution of both H0 ($\mu = 0$, $\sigma = 1$) and H1 ($\mu = 5$, $\sigma = 1$). The right hand limit to the acceptance region of H0 is marked. The question we ask is what proportion of H0's acceptance region overlaps with the distribution of H1. We will test the hypothesis with a sample size of $n = 10$ (since $\sigma = 1$, then $\sigma_{\bar{Y}} = 1/\sqrt{10} = 0.32$).

We follow these steps: first, we compute a z score by subtracting 5 (H1's μ) from 1.96, the cut-off point of H0's acceptance region, and divide by the standard error of the means (0.32). Then we look for the area beyond this z

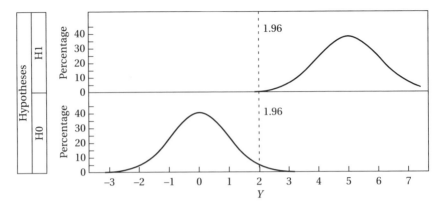

Figure 5.3. Distributions of H0 and H1. The mean of H0 is 0 and the mean of H1 is 5. The standard deviation of both distributions is 1.

score (column C). This is the area of the acceptance region of H0 which overlaps with H1's distribution. Thus:

$$z = \frac{5 - 1.96}{0.32} = 9.5$$

The area beyond this z score is approximately 0. This is the probability of committing a type II error and therefore the power of this test is approximately 1. Power can never be equal to 1, and the probability of committing a type II error can never be equal to 0. This is so because distributions of populations approach the horizontal axis asymptotically, and extend infinitely.

Practice problems 5.2

Compute β and $1 - \beta$ in the following examples. We will keep the same H0 ($\mu = 0$, $\sigma = 1$), but alter H1. In all cases, our sample will have a size of $n = 20$, so

$$\sigma_{\bar{Y}} = \frac{1}{\sqrt{20}} = 0.22$$

1. H0: $\mu = 0$, $\sigma = 1$; H1: $\mu = 3$, $\sigma = 1$; $n = 20$, $\sigma_{\bar{Y}} = 0.22$

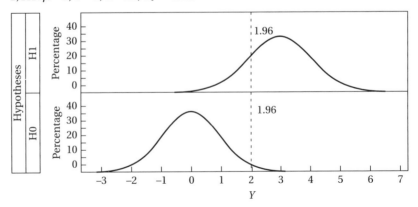

In this case $z = \dfrac{3 - 1.96}{0.22} = 4.73$. The area beyond this z score is approximately 0. This is the probability of committing a type II error, and therefore the power of this test is approximately 1.

2. H0: $\mu = 0$, $\sigma = 1$; H1: $\mu = 2$, $\sigma = 1$; $n = 20$, $\sigma_{\bar{Y}} = 0.22$

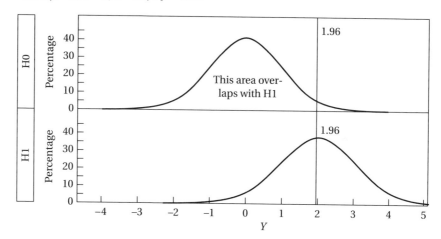

For this example $z = \dfrac{2 - 1.96}{0.22} = 0.18$. The area beyond a z score of 0.18 is 0.4286. This is the probability of committing a type II error, and therefore the power of this test is $1 - 0.4286 = 0.5714$. Clearly, the less different the two populations, the less powerful the test.

An interesting aspect of statistical power is that it can be increased by increasing the sample size. Thus, if instead of obtaining a sample of size $n = 20$ in the examples above, we obtained one of $n = 40$, we would have increased our power while decreasing our probability of a type II error. Thus: $\sigma_{\bar{Y}} = \dfrac{1}{\sqrt{40}} = 0.16$.

In the first example, we had H0: $\mu = 0$, $\sigma = 1$; H1: $\mu = 3$, $\sigma = 1$. Thus $z = \dfrac{3 - 1.96}{0.16} = 6.5$. The area beyond this z score is even closer to 0 than that of the z score computed with $\sigma_{\bar{Y}} = 0.22$. Thus, the power of this test is very near 1.

In the second example we had H0: $\mu = 0$, $\sigma = 1$; H1: $\mu = 2$, $\sigma = 1$. Thus $z = \dfrac{2 - 1.96}{0.16} = 0.25$. The area beyond this z score is 0.4013, a smaller probability of committing a type II error than that computed with a sample size of $n = 20$. The power of the test is now $1 - 0.4013 = 0.5987$ instead of 0.5714.

5.3 Examples of hypothesis tests using *z* scores

As was mentioned previously, although the null hypothesis is tested, the alternative hypothesis is the one usually generated by the scientist's research interest. Thus, whereas the researcher may want to show that there is a treatment affecting the behavior of his sample, the null hypothesis will propose that there is none. If H0 is rejected, then H1 is accepted as the most likely explanation: the sample supports that there is a treatment effect. The following examples are of pedagogical value only, and do not reflect actual data. Their purpose is to illustrate hypothesis testing with z scores.

1. An anthropologist is interested in studying the effect of religious affiliation on fertility. She works in a community which consists of a distinct religious group, and suspects that in this community large families are valued more than they are in the wider national community. The researcher wishes to determine if the small community has a significantly different mean family size from that of the national society. The anthropologist collects national census data and finds out that for the entire country the mean number of children per household is $\mu = 4$ with $\sigma = 1.8$. Thus H0:$\mu = 4$ and H1: $\mu \neq 4$. She decides to use a 0.05 level for α. The investigator obtains a sample of size $n = 10$ families and computes a sample mean of $\bar{Y} = 6$. Let us test the null hypothesis with a z score. Thus:

$$\sigma_{\bar{Y}} = \frac{1.8}{\sqrt{10}} = 0.57 \quad \text{and} \quad z = \frac{6 - 4}{0.57} = 3.5$$

The probability of finding this or a more extreme value is given by column C, and it is 0.0002. Since the probability is less than 0.05 (or, since the z score is larger than 1.96, the 0.05 cut-off point in the normal distribution table), the null hypothesis is rejected. The researcher also knows that she could be making a type I error, but the probability of that is rather low: 0.0002.

2. A researcher is interested in studying if a particular horticultural group consumes different numbers of animal species seasonally. To this effect, the anthropologist divides the year into the wet and dry seasons, and uses the group's diet of the wet season to generate the null hypothesis. The investigator determines that the mean number of animal species consumed during the wet season is $\mu = 10$ with a standard deviation of $\sigma = 3$. He returns to the community during the dry season and obtains samples from 20 households, and finds out that the mean number of animal species consumed during the dry season is $\bar{Y} = 9$. Therefore: H0: $\mu = 10$; H1: $\mu \neq 10$. The null hypothesis will be rejected at the usual $\alpha = 0.05$. Thus:

$$\sigma_{\bar{Y}} = \frac{3}{\sqrt{20}} = 0.67 \quad \text{and} \quad z = \frac{10 - 9}{0.67} = 1.49$$

In this example the researcher fails to reject the null hypothesis, and accepts that this population's number of animal species is stable through the year.

3. A team of anthropologists is interested in establishing if, in the community of their study, males and females have a different number of words to refer to soil quality (for example, the mineral content). It is known that, for the male population, the mean number of words for soil quality is $\mu = 4$ with a standard deviation of $\sigma = 1.1$. This information will be used as the null hypothesis' population. A female anthropologist takes a sample of $n = 30$

females and determines that the mean number of words for soil quality is $\bar{Y}=7$. Therefore: H0: $\mu=4$, $\sigma=1.1$; H1: $\mu\neq4$. The null hypothesis will be rejected at the usual $\alpha=0.05$. Thus:

$$\sigma_{\bar{Y}}=\frac{1.1}{\sqrt{30}}=0.20 \quad\text{and}\quad z=\frac{7-4}{0.20}=15$$

This is a highly significant result, which leads to the rejection of the H0 with a high confidence that the decision is not in error. The probability that a type I error was committed is close to 0.

5.4 One- and two-tail hypothesis tests

The reader has probably noticed that, up to this point, whereas all our null hypotheses have been stated exactly (say H0: $\mu=20$) our alternative hypotheses have not (H1: $\mu\neq20$) (except for the case when we illustrated the computation of β and $1-\beta$). In two-tail or two-way hypothesis testing, the researcher does not indicate in what direction the treatment could affect the data. Thus, the researcher does not say that she suspects (for example) that members of a certain religious group are likely to have larger or smaller family sizes. Two-tail tests are customary and widely accepted. Indeed, researchers rarely mention in writing that a test was two-tailed.

There are some situations which call for a one-tailed approach however. One-tailed tests are directional, that is, they state in what direction the treatment is expected to affect the data. Notice that both the H0 and the H1 are changed from what we did previously:

Two-tailed	*One-tailed*
	For a positive treatment effect:
H0: $\mu=4$ and H1: $\mu\neq4$.	H0: $\mu\leq4$ and H1: $\mu>4$
	For a negative treatment effect:
H0: $\mu=4$ and H1: $\mu\neq4$.	H0: $\mu\geq4$ and H1: $\mu<4$

Also notice that whereas the alternative hypothesis states that the mean will be greater or lesser than H0's , the null hypothesis states that the mean could be affected in an opposite manner from that stated by H1, or that it could be not affected at all. In other words, there are three possibilities which need to be covered in a one-tailed test: the direction stated by H1, the opposite direction, and no direction at all. The last two possibilities are stated in the H0.

One-tail tests may be used if the researcher either knows that the treatment can not affect the data in any but the direction stated by H1, or if he is simply not interested in any other direction. Conceivably, if a drug is tested in patients, the researchers may truly know that the drug will only improve the patient's health.

The main reason one-tailed tests are not frequently performed (so that researchers say in publications that they applied a one-instead of a two-tailed test) is that such tests are associated with a greater probability of a type I error, even if the same α level is used. If the investigator is going to reject H0 by looking at one tail only and wishes to use $\alpha = 0.05$, then she needs a cut-off point of H0's distribution beyond which all of the 0.05% of the distribution is found. This cut-off point is 1.645, which is actually not found in our table, so we use 1.65 instead. Previously, we rejected the H0 if the z score was greater than or equal to 1.96, which has 0.025 of the distribution beyond it. In a one-tailed test however, we can reject the null hypothesis on one side only, so we use a cut-off point with 0.05 beyond it.

The difference between these two approaches is best illustrated with an example. Let us say that an anthropologist has been working in a country in which the national mean number of children produced by women at the end of their reproductive career is $\mu = 6$ with a $\sigma = 2$. The researcher is interested in determining if women who have had higher education have a lower fertility. The researcher chooses to use a one-tailed test because he feels there is strong-enough evidence to suggest that higher education lowers, but never increases, the fertility of women. The anthropologist then takes a sample of size $n = 30$ educated females and determines that the mean number of children produced by these women at the end of their reproductive careers is $\overline{Y} = 5.4$. The researcher decides to use an $\alpha = 0.05$, so the cut-off point will be 1.65 instead of 1.96. The hypotheses are stated as follows: H0: $\mu \geq 6$ and H1: $\mu < 6$. A z score is computed as done previously:

$$\sigma_{\overline{Y}} = \frac{2}{\sqrt{30}} = 0.36 \quad z = \frac{5.4 - 6}{0.36} = -1.67$$

Since the z score is greater in absolute value than 1.65, the null hypothesis is rejected. If the researcher had decided to use a two-tailed test he would not have rejected H0 because the z score is less than 1.96.

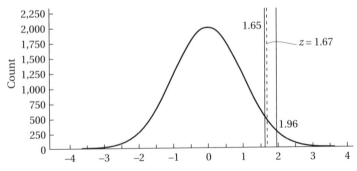

Figure 5.4. A z score (1.67) which would not be significant in a two-tail test but is significant in a one-tail test.

5.5 Assumptions of statistical tests

Although this chapter has covered the topic of hypothesis testing in general, a few words should be said about statistical tests of hypothesis in particular. When a hypothesis is to be tested with a statistical test, the test requires a number of assumptions from the data and the manner in which the data were collected. Some of these assumptions are specific to the particular test, so they will be discussed in the appropriate chapter. Some however, are shared by many tests, so they are discussed here:

1. Random sampling. All tests assume that the sample was collected giving each member of the population an equal chance of being selected. Only with a random sample do we have an accurate representation of the population.

2. Independence of variates. This is an assumption which is met by a sound sampling design. The assumption means that, if an observation is collected, it will not influence which observation is collected next. Thus, the variates are independent from each other. Anthropologists who work with historical demographic data may confront a source of data which does not meet this assumption. If the researcher collects the number of deaths per month in a community, and the community is affected by a cyclical mortality pattern, then the number of deaths in a month is not independent from the number of deaths in the next month. For example, the datum of a month in the rainy season will have a high correlation with the data of the other months of the rainy season. In this case, statistical analyses that are designed to deal with auto-correlated data must be applied (Madrigal, 1994).

3. The data are normal. As was mentioned previously, it is known that many continuous numerical variables **are** normally distributed. Thus, it is known that the principles about hypothesis testing which apply to the normal distribution also apply to (e.g.) height and many other variables. In other words, we can safely say that 68.26% of all subjects are found within the first positive and the first negative standard deviations, that 95.44% are found within the first two positive and negative standard deviations, etc. Therefore, we can also use the principles of hypothesis testing already demonstrated in this chapter, and decide that if an outcome is found within 95% of the distribution it will not be considered to be *statistically significantly* different from the mean. If the outcome is found in the 'left-over' 5%, then we can decide that it is statistically significantly different from the mean. The point here is that if a population is normally distributed, we can use these principles to test hypotheses about the population.

What if the population is not normally distributed? Then we can simply not use statistical tests that assume normality. If we did, we could be rejecting null hypothesis in error, not because of the type I and type II errors implicit in every statistical test, but because the population does not follow a normal distribution. The assumption of normality is frequently not given much importance because most samples of size $n = 30$ **are** normally distributed. However, SAS offers an extremely easy way of testing if data are normally distributed. Thus, for all tests which assume data to be normal, we will test the normality of the data before applying the test.

If a data set is found to be non-normal, then the researcher has two options open:

1. A non-parametric test is applied. Non-parametric tests have few assumptions, are easy to compute and can be applied to small data sets. The negative aspect about non-parametric tests is that they are less powerful (statistically speaking) than are parametric ones. This book covers non-parametric statistical tests which can be applied instead of the tests which assume normality.

2. The data set can be transformed, so that it does become normal. Fortunately, this is an exceedingly easy task with most statistical computer packages. The purpose of transforming the data is not for the researcher to 'prove' his hypothesis, but simply to test a hypothesis with a particular statistical test. The outcome of the test is not under the control of the researcher. Students are frequently uneasy about the notion of transforming a data set. However, transforming a data set is what is done when a researcher changes a series originally measured in inches into centimeters. There is no fundamental scientific reason why a data set must be measured in a particular way (the common linear scale to which we are accustomed). The same data can be measured in another scale which also makes the data normal. Commonly used transformations are: increasing the power, taking the square root, or obtaining the logarithm of the original variable. The logarithmic transformation is frequently used in archaeology (see, for example, Reitz et al., 1987).

5.6 Hypothesis testing with the *t* distribution

In order to introduce the *t* distribution, it is worthwhile reviewing what we have covered so far in terms of hypothesis testing. We have tested a null hypothesis that proposes that a sample mean $\overline{Y}$ belongs to a population whose parametric mean is μ. The null hypothesis was tested by computing a z score as:

$$z = \frac{Y - \mu}{\sigma_{\overline{Y}}}, \quad \text{where } \sigma_{\overline{Y}} = \frac{\sigma}{\sqrt{n}}$$

Notice that the hypothetical value of μ *is given by the researcher*. For example, we proposed that if a horticultural group did not experience seasonal variation in the number of animal species it consumed during the entire year, then the sample means obtained during different times of the year were the same, because they sampled the same population and thus had the same μ. Thus, it is proposed that $\overline{Y}_{\text{wet season}} = \overline{Y}_{\text{dry season}} = \mu$. We also tested the hypothesis that, in a particular society, the number of words for soil quality used by males and females was the same. We simply proposed that the parametric number of words (μ) for females was the same as that for males. Thus, a researcher can always propose a hypothetical value for μ, and test if the obtained sample or observation differs from it. But in all these examples we assumed knowledge of σ, knowledge we do not usually have. Therefore, we need another means to test hypotheses 'in the real world', that is, without knowledge of the population's σ. Our needs are met by computing *t* scores instead of *z* scores, and by using the *t* distribution, instead of the normal distribution.

Just as we can transform any sample mean into a *z* score, we can transform any sample mean into a *t* score. The difference is that we use the sample formula for computing the standard error of the means. The formula for the *t* score follows:

Formula 5.1 A sample mean's *t* score

$$t = \frac{\overline{Y} - \mu}{s_{\overline{Y}}}, \quad \text{where}$$

$$s_{\overline{Y}} = \frac{s}{\sqrt{n}}$$

The main difference between a *t* score and a *z* score is that the former computes the standard deviation using the sample instead of the population formula. That is, the former divides the sum of squares by the degrees of freedom ($n-1$) instead of N. Since the standard deviation is computed according to a sample's degrees of freedom (df), then it follows that if a sample is to be transformed into a sample of *t* scores (just like when a sample is transformed into a normal distribution), the shape of the *t* distribution will depend on the specific df of the specific sample. Therefore, there is not one *t* distribution, but an infinity of them, from df $= 1$ to df $= \infty$ (infinity). The larger the value of df, the more similar the *t* distribution is to the normal distribution (see figure 5.5). If $n = 30$, then the *t* distribution is virtually the same as the normal one and, at $n = \infty$, the *t* distribution is the normal distribution. For smaller samples, the *t* distribution is flatter at the center, having more variates at the tails. Table 2 in Appendix C shows the critical values of the *t* distribution.

The *t* distribution is used for hypothesis testing just like the normal distribution. That is, we compute a *t* score and ask what the probability is of obtaining this or a more extreme score. We will continue to use the same alpha level, that is, if the probability associated with a particular score is less than or equal to 0.05, we will declare the score's difference from μ statistically significant. Because the shape of the *t* distribution is symmetrical, it does not matter if a *t* score is positive or negative. The *t* distribution also

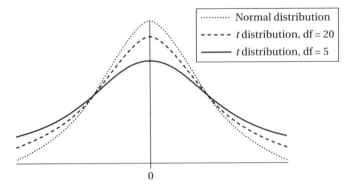

Figure 5.5. The normal and the *t* distribution (at different df).

asymptotically approaches the horizontal axis, so that we can never say that the probability associated with a very unusual score is 0; we can only say that such probability approaches 0.

Since there is a *t* distribution for each df from 1 to ∞, the cut-off point or **critical value** (cv) between the distribution's acceptance and rejection areas is specific to each df's distribution. Therefore, whereas with the normal distribution we knew that a *z* score whose value was 1.96 had a 0.05 probability of occurrence, and we used it as a cut-off point, there is a different cut-off point for each *t* distribution. Thus, if we wish to determine if a *t* score is significantly different from the parametric mean, we need to look at the table, having established the significance level and the df. In case a one-tail hypothesis is being proposed, we need to use a critical value which has all of the distribution's 5% on its side, as opposed to having 2.5% equally distributed on each tail. Thus, for a one-tail hypothesis, we would use the 0.1 (0.05 + 0.05 = 0.1) column. However, two-tail hypotheses are usually preferred (a smaller *t* value would be considered to be significant in a one-tail test as opposed to a two-tail test).

Let us review the steps involved in hypothesis testing using the *t* distribution.

1. State the null (H0) and alternative (H1) hypothesis. The null hypothesis will usually be of the sort H0: μ = a hypothetical number.
2. Establish the level of statistical significance.
3. Collect the sample.
4. Compare the sample with the null hypothesis, and reach a conclusion about which hypothesis to accept. The comparison is done by computing a *t* score as done in formula 5.1.

5.7 Examples of hypothesis tests with *t* scores

We use here the same examples we used in section 5.3, where we tested these hypotheses with *z* scores. However, since we use here the sample formula to estimate s instead of the population formula to estimate σ, the values of $s_{\bar{Y}}$ and $\sigma_{\bar{Y}}$ are different.

1. In this example, the researcher is interested in investigating whether a religious isolate has a significantly different mean number of children per household from that of the national society. The investigator obtains a sample of size $n = 10$ families, whose $\bar{Y} = 6$ and $s = 2$. The researcher uses census information to propose the null hypothesis' μ. For the entire country, the mean number of children per household is 4. The H0 is $\mu = 4$. The H1 is $\mu \neq 4$; thus, the test is two-tailed. An α level of 0.05 will be used. The t score is computed as follows:

$$s_{\bar{Y}} = \frac{2}{\sqrt{10}} = 0.63 \quad \text{and} \quad t = \frac{6-4}{0.63} = 3.17.$$

At $\alpha = 0.05$ for df $= 10 - 1 = 9$, the critical value is 2.262. Thus, we reject H0 at a 0.05 level. As a matter of fact, our t statistic is greater than the critical value at 0.02 % (cv $= 2.821$), giving us more confidence that we are not committing a type I error. Since our t statistic is smaller than the critical value at 0.01 (cv $= 3.25$) we can say that the probability of obtaining this t score, if the null hypothesis is true, lies somewhere between 0.01 to 0.02.

2. The second example dealt with a horticultural group whose seasonality of diet was being studied. The anthropologist uses the mean number of animal species consumed during the wet season (10) to propose the μ of the null hypothesis. Thus, H0: $\mu = 10$ and H1: $\mu \neq 10$. As usual, a two-tailed test at a 0.05 level will be used.

The researcher investigates $n = 20$ households, and determines that the mean number of animal species consumed during the dry seasons was $\bar{Y} = 9$ with a standard deviation $s = 3.5$. The t score is computed as follows:

$$s_{\bar{Y}} = \frac{3.5}{\sqrt{20}} = 0.78 \quad \text{and} \quad t = \frac{10-9}{0.78} = 1.28.$$

At $\alpha = 0.05$ with df $= 20 - 1 = 19$, the critical value is 2.093. Thus, the investigator fails to reject the null hypothesis: the community does not appear to consume a significantly different number of animal species during the dry season from the number consumed during the wet season.

3. The last example concerns the number of words to refer to soil quality by males and females in a community. The parametric mean of the null hypothesis is proposed based on the number of words used by males. Thus, H0: $\mu = 4$ and H1: $\mu \neq 4$. A sample of size $n = 30$ females is researched, and its mean number of words is determined to be $\bar{Y} = 7$, with a standard deviation $s = 1.8$. A two-tail test with an α level of 0.05 will be used. The t score is computed as follows:

$$s_{\bar{Y}} = \frac{1.8}{\sqrt{30}} = 0.33 \quad \text{and} \quad t = \frac{7-4}{0.33} = 9.1.$$

For df $= 30 - 1 = 29$ and at a level of 0.05 the critical value is 2.045. In fact, the t score is much greater than the critical value at 0.001 (3.659). Thus, the null hypothesis is rejected, and the alternative hypothesis that males and females in this community use different number of words for soil quality is accepted. What is the probability of committing a type I error in this situation? Less than 0.0001.

5.8 Reporting hypothesis tests

Different disciplines, and different journals within the same discipline, have different standards on how results of statistical tests should be reported. However, there is a minimum of information that should always be included when reporting statistical results: the value of the test statistic (say the t score), the degrees of freedom or the sample size, the probability associated with such a statistic (better known as the p value), and the conclusion reached about the hypothesis. How this information is written depends on the writer and the journal or book. For example, the researcher may report the results of the third example of the previous section as $t = 9.1$, df $= 29$, $p < 0.0001$, or as $t_{(9)} = 9.1$, $p < 0.0001$, or as $t_{(df = 9)} = 9.1$***. The latter example illustrates the common use of asterisks, in which * $= 0.05$ (but not significant to the 0.01 level), ** $= 0.01$ (but not significant to the 0.0001 level) and *** $= 0.0001$. Another frequently used convention is that, if a t score (or an F ratio, as we will see in chapter 7) is less than 1 in value, it is always not significant, no matter the degrees of freedom. In that situation, researchers simply report, say, $t = 0.8$, ns (for non-significant).

If the statistical analysis is done with computers, then the exact probability of a statistic such as a t score is known. That is, instead of knowing that the probability associated with a specific score is, say, between 0.05 and 0.01, it is known that the probability is exactly, say, 0.025. In that case, researchers use an equals sign (instead of a '<' sign) when reporting the probability. The use of computers has simplified greatly what can be a laborious and error-prone task: looking up in a statistical table to find out if the results are significant. Since the computer prints the statistic, the degrees of freedom, and the p value associated with the statistic, there is no need to use the statistical table: if the probability is less than or equal to 0.05, then the statistic is significant.

Whatever specific notation researchers use, they must at some point specify:
- what hypothesis is being tested
- what statistical test is being used
- the results of the statistical test
- the conclusion reached about the hypothesis.

5.9 Key concepts

The principles of hypothesis testing
Alternative and null hypotheses
Significance level
Type I error
Type II error
Statistical power
One- and two-tail hypothesis tests
Independence of variates
Non-parametric test
t distribution

5.10 Exercises

1. Create a research problem of interest to you and describe your null and alternative hypotheses.
2. What is the difference between a t score and a z score?
3. Given a hypothesis test, compute the following.
    (a) If H0: $\mu=0$, H1: $\mu=2$, the standard deviation is $\sigma=1$, and $n=5$, what is the test's power?
    (b) If H0: $\mu=0$, H1: $\mu=2$, the standard deviation is $\sigma=1$, and $n=20$, what is the test's power?
    (c) If H0: $\mu=0$, H1: $\mu=3$, the standard deviation is $\sigma=1$, and $n=5$, what is the test's power?
    (d) If H0: $\mu=0$, H1: $\mu=3$, the standard deviation is $\sigma=1$, and $n=20$, what is the test's power?
4. Test the hypotheses that these sample means belong to the population proposed by the H0. Use both a z score and a t score.
    (a) For the z score: H0: $\mu=18$, $\sigma=2$, $n=90$, $\overline{Y}=20$. For the t score: H0: $\mu=18$, $s=2.3$, $n=90$, $\overline{Y}=20$.
    (b) For the z score: H0: $\mu=178$, $\sigma=15$, $n=300$, $\overline{Y}=172$. For the t score: H0: $\mu=178$, $s=20$, $n=300$, $\overline{Y}=172$.
5. Find out how statistical tests are usually reported in a journal in your field of study.

6 The difference between two means

In the last chapter we covered the topic of hypothesis testing with z and t scores. The hypothesis we tested in both cases was that a particular sample $\overline{Y}$ was obtained from a population with mean μ. We discussed that we could use z scores only if we knew the population's σ (a very rare situation), and that in most cases we use a t score which is computed using the sample's standard deviation. Notice that in either case we are comparing a sample mean with a hypothetical population. This is not a very common experimental design. Instead, it is more frequent to compare two samples with each other, testing the null hypothesis that both were collected from the same population. Such hypothesis is covered here, specifically the following cases:

1. the difference between two independent samples (e.g. the weight of two samples of children independently collected in two different locales of a community),
2. the comparison of an observation with a sample (e.g. an arrowhead with a sample of arrowheads)
3. the comparison of two paired groups (e.g. the weight of a group of individuals before and after a smoking-cessation program).

The reader should note that, in principle, all of these tests could be done using z scores if we had knowledge of the population parameters. Since this is rarely the case, we test these hypotheses using t scores.

6.1 The un-paired t test

This is a popular and frequently used statistical test, commonly referred to as 'the t test'. In reality, this test is a special case of an analysis of variance (ANOVA). However, the test is so popular that it is usually discussed separately from that of ANOVA. The test is used if a researcher needs to compare two samples for significant differences, such as the mean age at marriage of females of two ethnic groups, or the mean length of projectile points, or the mean height of two groups of children, etc. The null and alternative hypotheses in a two-tail test are H0: $\mu_1 = \mu_2$, and H1: $\mu_1 \neq \mu_2$. A one-tail test should only be done (as always) when the researcher knows, or is interested in testing, only that one group has a greater or smaller mean than the other. In a one-tail test, the null and alternative hypothesis are H0: $\mu_1 \geq \mu_2$ and H1: $\mu_1 < \mu_2$, or H0: $\mu_1 \leq \mu_2$ and H1: $\mu_1 > \mu_2$, depending on the direction of the treatment effect.

The problem we face in this situation is that we need to agree how different two means have to be for us to decide that they were not sampled from the same population. As we learned when we discussed the distribution of sample means, means of size *n* will be normally distributed around the parametric mean. Thus, some means from the same population may be different enough from each other for us to decide (in error) that they were not sampled from the same population. If two means do not differ at all, their difference is 0, which is the mean of the distribution of sample-mean *differences*. The more different the two samples are, the larger (negative or positive) the difference between the means will be. At the same time, as the difference increases in value it will be farther apart from the mean of sample differences, and it will have a lower frequency because it will be located toward the tail of the distribution. What we are discussing here is nothing more than a normal distribution, whose properties we are familiar with. If a difference falls within the customary 95% acceptance region, it will be declared to be non-significant, and the null hypothesis will be accepted. If the difference falls in the rejection region, usually distributed equally in both tails, then the difference will be declared to be significant, and the null hypothesis will be rejected. This rejection will be made with the acknowledgment that such a decision could be in error, accompanied by the probability (*p* value) that we are mistaken.

The formula for the *t* score to test the null hypothesis that both samples were obtained from the same population is very similar to that of the *t* score we used in the last chapter (to test that a sample was obtained from a population of mean μ). Both have in the numerator a difference, which in the *t* score we used in the last chapter was the difference between the sample mean $\overline{Y}$ and the population mean μ. In the *t* score we will use in this chapter, the difference is between that of the two $\overline{Y}$s and that of the two μs $([\overline{Y}_1 - \overline{Y}_2] - [\mu_1 - \mu_2])$. Since, according to the null hypothesis, the two samples were obtained from the same population, the difference between the two parametric means is 0. Thus, the second term $(\mu_1 - \mu_2)$ is removed, and the numerator is simply $(\overline{Y}_1 - \overline{Y}_2)$. The denominator of the *t* test to compare two samples is also very similar to that of the *t* score to compare a sample to a population. The reader should recall that in the latter case we obtained the denominator by dividing the sample's standard deviation by the square root of the sample size. We did this because the standard deviation of sample means depends on the sample size as well as on the variation of items in the sample. In the *t* score we will use in this chapter we need a similar measure, but one that reflects not the standard error of means, but the standard error of *mean differences*. The standard error of mean differences $s_{\bar{Y}_1 - \bar{Y}_2}$ also incorporates the sample sizes (n_1 and n_2), and standard deviations (actually, the variances s_1^2 and s_2^2 are chosen for ease of computation). The formula presented below looks cumbersome, but has the advantage that it can be applied to any two samples, even if their sizes are different. When performing hand calculations, the researcher should try to keep as many digits as is possible, rounding only at the end. That way rounding error does not accumulate through a series of calculations.

Formula 6.1 The _t_ score to test that two sample means were obtained from the same population

$$t = \frac{\bar{Y}_2 - \bar{Y}_2}{\sqrt{\left[\frac{(n_1 - 1)s_1^2 + (n_2 - 1)s_2^2}{n_1 + n_2}\right]\left(\frac{n_1 + n_2}{(n_1)(n_2)}\right)}}$$

with df $= n_1 + n_2$

Let us practice the computation of the _t_ test with the following data set: is there a significant difference in the mean length of projectile points (in centimeters) collected in these two fictitious sites?

Site 1	Site 2
4.5	5.2
5.2	5.7
4.3	6.0
4.7	6.7
4.0	5.5
3.9	5.4
5.8	6.8
2.8	

We have: $n_1 = 8$, $n_2 = 7$, $\bar{Y}_1 = 4.4$, $\bar{Y}_2 = 5.9$, $s_1^2 = 0.81$, $s_2^2 = 0.4$.

We follow the usual steps when testing a hypothesis.

1. First, we state the null and alternative hypothesis: H0: $\mu_1 = \mu_2$, H1: $\mu_1 \neq \mu_2$.
2. The alpha level will be 0.05. For the appropriate degrees-of-freedom ($7 + 8 - 2 = 13$) the critical value is 2.16. Thus, if our _t_ score is greater than (in absolute value, since the _t_ distribution is symmetrical) or equal to 2.16, we will reject the H0.
3. The sample has been collected and given to us.
4. The _t_ score is computed and a decision is reached.

$$t = \frac{4.4 - 5.9}{\sqrt{\left[\frac{(8 - 1)0.81 + (7 - 1)0.4}{8 + 7 - 2}\right]\left(\frac{8 + 7}{(8)(7)}\right)}}$$

$$= \frac{-1.5}{\sqrt{\left[\frac{8.07}{13}\right]\left(\frac{15}{56}\right)}} = \frac{-1.5}{\sqrt{[0.62](0.21)}} = \frac{-1.5}{0.41} = -3.66.$$

The *t* score is greater than the critical value, so we reject at the 0.05 level. In fact, our *t* statistic lies somewhere between the critical values of $= 0.01$ and 0.001. We report our results as: $t = -3.67$, $df = 13$, $0.001 > p < 0.01$.

Practice problem 6.1

The following fictitious data are the number of surviving children of females of age > 50 in two villages. Do females in the two villages have significantly different numbers of children?

Group 1	Group 2
4	4
3	8
2	10
5	5
1	1
0	0
7	3
8	6
4	9

We have: $n_1 = n_2 = 9$, $\bar{Y}_1 = 3.78$, $\bar{Y}_2 = 5.11$, $s_1{}^2 = 6.94$, $s_2{}^2 = 12.11$. We follow the usual steps.

1. First, we state the null and alternative hypotheses: H0: $\mu_1 = \mu_2$, H1: $\mu_1 \neq \mu_2$.

2. The alpha level will be 0.05. For the appropriate degrees of freedom $(9 + 9 - 2 = 16)$ the critical value is 2.12.

3. The sample has been collected and given to us.

4. The *t* score is computed and a decision is reached.

$$t = \frac{3.78 - 5.11}{\sqrt{\left[\frac{(9-1)6.94 + (9-1)12.11}{9+9-2}\right]\left(\frac{9+9}{(9)(9)}\right)}}$$

$$= \frac{-1.33}{\sqrt{\left[\frac{152.4}{16}\right]\left(\frac{18}{81}\right)}} = \frac{-1.33}{\sqrt{[9.525](0.22)}} = \frac{-1.33}{2.0955} = -0.63$$

Our decision is to accept the null hypothesis, since the *t* statistic is smaller than the critical value. Anyway, because the statistic is less than 1, we need not consult the table, since a *t* statistic smaller than 1 is always non-significant. We report our results as: $t = -0.92$, $df = 16$, ns.

An example of a *t* test using SAS ASSIST

The archaeological data set used in the example above was entered in two columns (the first had the length, the second the site). If the observations of site 2 had been entered before those of site 1, the data would have had to be sorted. From the primary menu, the following path was used: 1. data analysis, 2. ANOVA, 3. *t* tests, 4. comparison of two means. Site was chosen as the classification variable, length as the dependent variable. The titles were added. The output is reproduced below:

```
                        t test of fictitious data
                Length (cm) of projectile points in two sites

                            T TEST PROCEDURE
        Variable: LENGTH
        SITE   N    Mean          Std Dev       Std Error    Minimum       Maximum
        ------------------------------------------------------------------------
        1      8    4.40000000    0.90079330    0.31847853   2.80000000    5.80000000
        2      7    5.90000000    0.63245553    0.23904572   5.20000000    6.80000000

        Variances   T          Method         DF     Prob>|T|
        -----------------------------------------------------
        Unequal    -3.7669    Satterthwaite   12.5   0.0025
                              Cochran                0.0078
        Equal      -3.6763                    13.0   0.0028←p value for H0

        For H0: Variances are equal, F'=2.03 DF=(7,6) Prob>F'=0.4072
```

Note that SAS prints items such as the sample size, the mean and standard deviation. At this point we still have not explained why SAS prints two *t* scores, next to the column headed 'variances'. Please notice that in the second row of the column T(equal), the *t* value is -3.67, the degrees of freedom are 13, and the *p* value is 0.0028. By simply looking at the *p* value, we reach our decision without consulting the *t* table: if the *p* value ≤ 0.05, we reject.

6.2 Assumptions of the un-paired *t* test

It is interesting that the assumptions of statistical tests tend to receive little attention in many texts, and that they are usually discussed after the test itself is discussed, even though the assumptions should be tested before the test is performed. Researchers should not proceed to apply a *t* test without testing some fundamental assumptions about the data, because if the data violate the assumptions, the results of the test may be meaningless.

In the last chapter we discussed several assumptions of many statistical tests. Thus, they will only be briefly mentioned here. The assumptions which must be tested to perform the *t* test will receive more attention.

1. Random sampling. The *t* test assumes that the samples were collected giving each member of the population an equal chance of being selected.

2. Independence of variates. The assumption means that, if an observation is collected, it will not influence which observation is collected next in the same sample. The assumption also extends to the other sample: both are totally independent from each other.

3. *The data are normal.* The assumption of normality is frequently not given much importance because most samples of size $n = 30$ *are* normally distributed. However, SAS offers an extremely easy way of testing if data are normally distributed: when requesting a univariate analysis of the data, request also that *each* sample be tested for normality. The output will print the W statistic which, if significant, means that the samples are not normal. In other words, the W statistic tests the hypothesis that the samples are normal. If the statistic is not significant, the null hypothesis that the data are normal is accepted, and the *t* test can be performed. Going this extra step is hardly back-breaking, because researchers would want to look at some summary statistics before doing the *t* test, anyway. PROC UNIVARIATE is the appropriate procedure but, as far as I could ascertain, this procedure is not accessible through the menu (the procedure for simple statistics available through the menu is PROC MEANS, which does not offer normality testing). Thus, the SAS code had to be written in the editor, and is reproduced below:

Testing the assumption of normality with SAS

```
data ttest1;
input length site;
cards;
4.5  1
5.2  1
4.3  1
4.7  1
4.0  1
3.9  1
5.8  1
2.8  1
5.2  2
5.7  2
6.0  2
6.7  2
5.5  2
5.4  2
6.8  2
;
proc univariate normal;
var length; by site;
```

The results of the analysis are reproduced below. Notice that the *p* values printed next to the W statistic is not equal to or less than 0.05. Therefore, we accept the null hypothesis that the data are normal, and can proceed with the computation of the t test. For the sake of space, the entire output is not reproduced. However, it should be noted that PROC UNIVARIATE provides more information about the samples than does PROC MEANS.

```
------------------------SITE = 1-----------------------
          Univariate Procedure
Variable = LENGTH
                 Moments
N               8          Sum Wgts     8
Mean            4.4        Sum          35.2
Std Dev         0.900793   Variance     0.811429
Skewness        -0.26112   Kurtosis     0.776582
USS             160.56     CSS          5.68
CV              20.47258   Std Mean         0.318479
T:Mean = 0      13.81569   Pr > |T|     0.0001
Num ^ = 0       8          Num > 0      8
M(Sign)         4          Pr >= |M|    0.0078
Sgn Rank        18         Pr >= |S|    0.0078
W:Normal        0.980257   Pr < W       0.9602←p value


------------------------SITE = 2-----------------------
          Univariate Procedure
Variable = LENGTH
                 Moments
N               7          Sum Wgts     7
Mean            5.9        Sum          41.3
Std Dev         0.632456   Variance     0.4
Skewness        0.647476   Kurtosis     -1.3295
USS             246.07     CSS          2.4
CV              10.71959   Std Mean     0.239046
T:Mean = 0      24.68147   Pr > |T|     0.0001
Num ^ = 0       7          Num > 0      7
M(Sign)         3.5        Pr >= |M|    0.0156
Sgn Rank        14         Pr >= |S|    0.0156
W:Normal        0.886582   Pr < W       0.2673←p value
```

4. *The variances are homogeneous.* The reader should recall that a t score that tests if two samples are significantly different has as its denominator the standard error of mean differences or $s_{\bar{Y}_1 - \bar{Y}_2}$. This statistic incorporates the variances and sizes of both samples, and in effect combines them. Since, according to the null hypothesis, both samples came from the same population, they should have homogeneous variances (a condition known as homoscedasticity). If the variances are not homogeneous (heteroscedasticity) then it is possible for a null hypothesis to be rejected not because the sample means truly differ from each other but because the sample variances do.

If using SAS/ASSIST, an investigator can test this assumption with great ease: the user would simply click the appropriate button in the t test menu (Approximate t statistic for unequal group variances), and SAS will: first test the null hypothesis that H0: $\sigma_1^2 = \sigma_2^2$, and then print two t scores, one to be used if the variances are homogeneous, and one to be used if the variances are not homogeneous. Using the previous data set, the test for homogeneity of variances is

```
For H0: Variances are equal, F'=2.03 DF=(7,6) Prob>F'=0.4072.
```

As always, if the *p* value is less than or equal to 0.05, the null hypothesis is rejected. Since the *p* value is 0.4072, the null hypothesis that the variances are homogeneous is accepted, and the *t* score for equal variances (*t* = − 3.6763) is used. If the variances had been heterogeneous, the other *t* score would have been used (*t* = −3.7669), and the researcher would have noted in writing that a test of equality of means of two samples whose variances are heterogeneous was used.

```
Variance   T          Method         DF    Prob>|T|
---------------------------------------------------------------
Unequal    −3.7669    Satterthwaite  12.5  0.0025 ← use if variances
                      Cochran         .    0.0078    heterogeneous
Equal      −3.6763                   13.0  0.0028 ← use if variances
                                                     homogeneous
```

Thus, if a researcher has access to SAS, the testing of the homoscedasticity assumption is simply part of the *t* test itself. If investigators do not have access to a computer package that does this, they must test the assumption by hand. A commonly used test is the Fmax test, in which the larger variance is divided by the smaller variance, and the ratio (known as the Fmax ratio) is compared with a table of critical values (table 3 in Appendix C). Two degrees of freedom are needed to use this table: the first is the number (*a*) of groups compared (*a*=2 in a *t* test), and the second is *n*−1, where *n* is the size of the smaller sample. If the Fmax statistic is greater than or equal to the critical value then the null hypothesis of homoscedasticity is rejected. In our example, the two variances are 0.811429 and 0.4. Thus, Fmax=0.8114/0.4=2.03 with df=2, 6. The critical value is 5.82, so the null hypothesis is accepted: the variances are homogeneous.

6.3 A research example of the un-paired *t* test

An example *t* test can be found in an article that discusses animal bones recovered from excavations at Chavín de Huántar, Peru (Miller and Burger, 1995). Miller and Burger indicate that the early embrace of camelid pastoralism by the people of Chavín de Huántar (Urabarriu phase) may have played an important role in their economy and aided in the tremendous influence of the Chavín cult. An attempt to identify the exact species present in the archaeological site is difficult since, morphologically speaking, the four camelid species do not differ. However, they do differ in size: among contemporary camelids guanacos are larger than llamas, llamas larger than alpacas, and alpacas larger than vicuñas. When measurements of the archaeological bones are plotted, they cluster into two distinct groups, one large and one small. Thus, apparently there were two species, but it is difficult to determine which of the two large (llama-guanaco), and which of the two small (alpaca-vicuña) species is present. Miller and Burger (1995) proceed to compare with

t-tests the phalanges of archaeological small and large camelids with the appropriate living species. On the basis of the unpaired t tests, as well as other diagnostic tests, Miller and Burger conclude that roughly 55% of the Urabarriu camelids were llamas and the remainder vicuñas (Miller and Burger, 1995).

6.4 The comparison of a single observation with the mean of a sample

Researchers may sometimes be in a situation in which they have a single variate (a projectile point, a bone, etc., symbolized as Y) of unknown origin. If they have one or more samples from which the observation could have come, then they can establish with a t score if the observation is or is not significantly different from the sample means. If the observation is not significantly different from the sample mean, the researchers have not of course proven that the observation belongs to the population from which the sample was obtained. But at least they know that the observation is not significantly different from the sample. The null hypothesis is that the μ of the population from which the observation was sampled is the same as the μ which is being estimated by the sample mean. Thus, H0: μ = the value estimated by the sample mean. The t score used to compare one observation from one sample is very similar to that presented in the previous section:

Formula 6.2 **The t score to compare a single observation with a sample mean.**

$$t = \frac{Y - \overline{Y}}{(s)\left[\sqrt{\dfrac{n+1}{n}}\right]}, \quad df = n - 1$$

Let us apply this formula by supposing that we found a projectile point of length = 5.5 close to two samples, whose means, standard deviations and sample sizes are:

```
SITE   n   Mean          Std Dev
------------------------------------
1      8   4.40000000    0.90079330
2      7   5.90000000    0.63245553
```

Let us compute a t score to compare our observation with both samples:

1. H0: μ = 4.4. $t = \dfrac{5.5 - 4.4}{(0.9)\left[\sqrt{\dfrac{8+1}{8}}\right]} = \dfrac{1.1}{(0.9)[1.06]} = 1.15, p > 0.05.$

2. H0: μ = 5.9. $t = \dfrac{5.5 - 5.9}{(0.63)\left[\sqrt{\dfrac{7+1}{7}}\right]} = \dfrac{-0.4}{(0.63)[1.07]} = 0.6, \text{ns}.$

In this example, the observation is not significantly different from either sample. At df = 7 the critical value is 2.365, and at df = 6 the critical value is 2.447.

6.5 The comparison of paired samples

One of the assumptions for the unpaired t test was that sampling be random. That is, both samples were obtained randomly from their populations. In a paired comparison, however, we frequently work with one sample instead of two, comparing the mean of the sample at two different times. Or, if we work with two distinct samples, they are paired or matched for variables that need to be controlled. Examples of this experimental design are the comparison of the mean weight in a group of individuals before and after a diet program, or the mean height in a group of children taken a year apart, or the achieved fertility in two groups of women matched for education, social class and contraceptive use but not matched for religion.

In a paired comparison, we can not use the t scores we used in the previous section. Even if we are working with two distinct samples (as in the example of the two groups of females), what interests us most is the magnitude of the differences between the two groups. Indeed, in a paired comparison we do not work with the data themselves, but instead create a new variable, the differences (D) between the observations of both groups. The null hypothesis states that the parametric mean of the population of differences from which our sample of differences was obtained has a value of 0. Thus: H0: $\mu_D = 0$. The alternative hypothesis is usually two-tailed. The structure of this t score is very similar to those previously discussed: in the numerator we are going to have a difference, namely, the difference between the mean sample difference ($\overline{D}$) and the difference between the parametric means of the populations from which our samples were obtained ($\mu_1 - \mu_2$). Since, according to the null hypothesis, the two populations are the same, $\mu_1 - \mu_2$ is 0, and the numerator consists only of $\overline{D}$. The denominator consists of a very familiar term, namely the standard error of the differences or $s_{\overline{D}} = \dfrac{s}{\sqrt{n}}$, where s is the standard deviation *of the differences* and n is the sample size. Thus, the formula is:

Formula 6.3 The t score for paired comparisons

$$t = \frac{\overline{D}}{s_{\overline{D}}}, \text{ where}$$

$$s_{\overline{D}} = \frac{s}{\sqrt{n}},$$

$$df = n - 1$$

Let us say that an anthropologist has been hired to determine if the test scores of a group of $n = 10$ children have changed significantly after the introduction of a new teaching methodology. The data are listed below, as well as a column of differences between both measures. It does not matter how the differences are obtained. However, if the computations are to be done by hand, then they should be computed to obtain the least number of negative differences. The usual steps in hypothesis testing are followed.

1. The null and alternative hypotheses are: H0: $\mu_D = 0$ and H1: $\mu_D \neq 0$.
2. An alpha level of 0.05 is chosen. At df = 9, the critical value is 2.262.
3. The data are collected:

Before	After	Difference (before – after)
80	95	−15
70	73	−3
60	55	5
70	75	−5
90	87	3
65	76	−11
70	65	5
55	50	5
80	82	−2
95	95	0

4. The t score is computed and a decision is reached.

$$\overline{D} = -1.8, \quad s_{\overline{D}} = \frac{7}{\sqrt{10}} = 2.2, \quad t = \frac{-1.8}{2.2} = 0.818$$

Since any t score whose value is less than 1 is known not to be significant, we accept the null hypothesis, and conclude that the new teaching methodology has not changed the test results. The results are reported as: $t = 0.818$, df = 9, ns.

Practice problem 6.3

An anthropologist interviews a group of males about their reproductive history. One question concerns how old the male was when he had his first child. In an attempt to test the reliability of the subjects' answers, the researcher re-interviews them three months later, hoping that if the subjects were not telling the truth the first time, they would have now forgotten what they told him the first time. Test the hypothesis that there is no difference in the answers.

1. The null and alternative hypotheses are: H0: $\mu_D = 0$ and H1: $\mu_D \neq 0$.

2. An alpha level of 0.05 is chosen. The critical value for df $= 14 - 1 = 13$ is 2.16.

3. The data are collected, and are listed below:

First interview	Second interview	Difference (first − second)
18	18	0
20	20	0
25	15	10
20	19	1
30	29	1
20	15	5
23	23	0
27	27	0
23	16	7
30	30	0
25	24	1
31	31	0
26	20	6
22	22	0

4. The *t* score is computed, and a decision is reached:

$$\overline{D}=2.21, \quad s=3.33, \quad s_{\overline{D}}=\frac{3.33}{\sqrt{14}}=0.89, \quad t=\frac{2.21}{0.89}=2.48, \quad df=14-1=13, \quad p<0.05.$$

The null hypothesis of no differences is rejected: the subjects provided significantly different answers to the questionnaire.

A paired *t* test in SAS/ASSIST

Let us do the example of the test scores in SAS. The data are entered in two columns, one having the first score, the other having the second score (notice that the data are entered differently for a paired and an un-paired *t* test). The path is: data analysis, ANOVA, *t* tests, paired comparisons. The output obtained is the following. Since the *p* value (0.4344) is greater than 0.05, the null hypothesis is accepted.

```
Analysis Variable : DIFF

Mean          Std Error    T             Probe>|T|
------------------------------------------------------
-1.8000000    2.2000000    -0.8181818    0.4344
------------------------------------------------------
```

6.6 Assumptions of the paired *t* test

The paired *t* test has the following assumptions.

   1. Random sampling. The paired *t* test assumes that the samples were collected giving each member of the population an equal chance of being selected.

   2. Independence of variates. The assumption means that, if an observation is collected, it will not influence which observation is collected next in the same sample. Of course, the two samples being compared are not independent from each other.

3. The data are normal. The assumption of normality affects the differences between the two measures, not the measures themselves. The assumption of normality can easily be tested by running a PROC UNIVARIATE test on the differences.

Testing the assumption of normality with SAS

```
data paired1;
input before after diff;
cards;
80   95   -15
70   73   -3
60   55    5
70   75   -5
90   87    3
65   76   -11
70   65    5
55   50    5
80   82   -2
95   95    0
     ;
proc univariate normal ; var diff;
```

The results of the analysis are reproduced below. The *p* value printed next to the *W* statistic is not equal to or less than 0.05. Therefore, we accept the null hypothesis that the data are normal, and can proceed with the computation of the *t* test. (By default, PROC UNIVARIATE tests the hypothesis that the mean of the sample is equal to 0, which happens to be the null hypothesis of the paired *t* test if a sample of differences is being analyzed. Thus, the *p* value printed below is the same as that obtained previously with the *t* test for paired samples. The *p* value is marked as 'H0 test').

Univariate Procedure

Variable = DIFF

	Moments			Quantiles (Def = 5)				
N	10	Sum Wgts	10	100%	Max	5	99%	5
Mean	−1.8	Sum	−18	75%	Q3	5	95%	5
Std Dev	6.957011	Variance	48.4	50%	Med	−1	90%	5
Skewness	−0.847	Kurtosis	−0.18523	25%	Q1	−5	10%	−13
USS	468	CSS	435.6	0%	Min	−15	5%	−15
CV	−386.501	Std Mean	2.2	1%	15			
T:Mean=0	**−0.81818**	**Pr>|T|**	**0.4344 ←H0 test**		Range	20		
Num ^= 0	9	Num>0	4		Q3−Q1	10		
M(Sign)	−0.5	Pr>=|M|	1.0000		Mode	5		
Sgn Rank	−3.5	Pr>=|S|	0.7031					
W:Normal	0.893596	Pr<W	0.1774 **←p value for normality**					

```
                        Extremes
          Lowest   Obs   Highest   Obs
            -15(    1)       0(      10)
            -11(    6)       3(       5)
             -5(    4)       5(       3)
             -3(    2)       5(       7)
             -2(    9)       5(       8)
```

6.7 A research example of the paired t test

An interesting use of a paired t test is presented by Neiman (1995). Neiman investigates the level of stylistic variation in ceramics, and attempts to estimate the parameter of diversity (θ) with two different statistics (t_F and t_E). Both statistics are computed for $n = 35$ Illinois Woodland assemblages, and are tested for significant differences. A paired t test was appropriate in this case because two different measures were taken in the same assemblage. Neiman (1995) states that 'The mean of the paired differences (t_F-t_E) was also computed and found to be statistically indistinguishable from 0 ($t = .002$, $p = .0098$)'. In a single sentence, much information was conveyed: the t statistic, its probability (the sample size had been mentioned previously, so there was no need to re-state the degrees of freedom), and the decision about the null hypothesis, as well as the hypothesis itself.

6.8 Key concepts

The use of z scores vs t scores. Why is it unlikely that you will use z scores in your research?

Unpaired t test: when is it appropriate to use?

Assumptions of the un-paired t test

Definitions of homoscedasticity and heteroscedasticity

The comparison of a single individual to a sample mean

Paired t test: when is it appropriate to use?

Assumptions of the paired t test

6.9 Exercises

1. A linguist studying two traditionally non-English-speaking communities wants to know which group of school children is scoring better on tests which are offered only in English. Below are the test scores on a recent exam offered in English:

Children of community 1	Children of community 2
65	70
59	80
65	70
59	80
90	85
91	78
87	94
100	69
77	89
69	75
89	100
92	61

2. The following data were kindly provided by White (work in progress). They are the lengths in millimeters of rangia shells (the left valve) from two different levels from the Depot Creek site in Northwest Florida. Only part of the data is shown. Test the hypothesis that the rangia shells from the two sites do not differ in their mean length. Test the assumptions of normality of each sample, and of homogeneity of variances.

Length	Level	Length	Level
32.6	1	41.4	1
32.6	1	42.1	1
34.1	1	41.9	1
33.0	1	42.6	1
30.2	1	41.7	1
33.0	1	39.9	1
32.4	1	43.0	1
30.9	1	40.6	1
31.6	1	32.3	2
31.1	1	32.8	2
32.5	1	33.1	2
33.0	1	35.2	2
32.8	1	36.8	2
32.8	1	32.7	2
33.9	1	31.2	2
35.3	1	33.3	2
31.5	1	33.6	2
32.2	1	42.7	2
34.5	1	34.4	2
30.0	1	33.8	2
42.5	1	33.4	2
37.4	1	33.6	2

Length	Level	Length	Level
31.6	2	32.0	2
34.2	2	31.6	2
34.5	2	33.0	2
33.5	2	35.1	2
35.8	2	29.5	2
36.2	2	33.9	2

3. An anthropologist is working with a community which is experiencing rapid acculturation. Particularly, many individuals are adopting a new supernatural belief system, in conjunction with their previous one. The anthropologist is interested in determining if the number of idols displayed in households changes as a result of such acculturation, in a period of a year. The researcher visits a number of households at the beginning of his fieldwork, and counts the number of idols displayed. A year later, the anthropologist visits the same households, and counts the number of idols. Has there been a significant difference? Test the assumption of normality of the differences.

OBS	Number of idols first visit	Number of idols second visit
1	6	8
2	1	5
3	0	1
4	3	3
5	5	7
6	8	9
7	7	7
8	6	4
9	10	10
10	2	3
11	6	5
12	6	8
13	8	9
14	5	6
15	0	1
16	9	10
17	4	8
18	2	2
20	4	8
21	8	9
22	4	3

4. A medical anthropologist wants to know if, within her sample, there is a significant decrease in the number of illness symptoms complained of by patients after they seek medical attention from a community-based traditional practitioner.

Before seeking medical advice	After seeking medical advice
10	8
8	3
7	2
4	6
5	0
3	1
2	0
8	4
3	4
6	3
9	5

Check the assumptions of a paired t test and ascertain whether there is a significant difference in symptoms reported after seeing a traditional healer.

7 Analysis of variance (ANOVA)

In the last chapter, we learned how to test whether two independent samples were obtained from the same population. In this chapter, we confront the same problem, but work with more than two samples.

This chapter provides a review of a limited aspect of ANOVA: model I, one-way analysis of variance. Excluded are the more complex experimental designs such as two-way, nested, and multi-way ANOVAS, as well as the computation of the added variance component of model II designs. These more complex ANOVAS are frequently discussed in psychology and biology texts, since in these fields carefully planned laboratory experiments are possible (see Sokal and Rohlf, 1981). This chapter begins by introducing the reader to a model I, one-way ANOVA, emphasizing the null hypothesis tested. The next section introduces the nomenclature, and explains how ANOVA information is usually displayed in tables. An example and a practice problem then follow. Finally a discussion of post-ANOVA comparison of means is presented.

7.1 One-way ANOVA

In a **one-way ANOVA** we deal with a problem similar to an unpaired t test, but use more than two samples. The null hypothesis is that the means we compare were obtained from the same population. In other words, the hypothesis states that the μ of the population from which the samples were obtained is the same. Thus H0: $\mu_1 = \mu_2 = \cdots = \mu_a$, where a is the number of samples being compared. The alternative hypothesis states that the sample means differ, and is generated by our research interest. It proposes that a treatment has affected at least one of the samples in such a way that the sample mean is unlikely to have been obtained from said population. 'Unlikely' is quantified with a probability statement. Therefore, we could compare the mean height of four samples of children of the same age obtained in four different neighborhoods and test the null hypothesis that the samples came from the same population. In religiously or ethnically diverse nations, we could test that more than two ethnic/religious groups differ in their mean age at menarche, or at marriage, or at birth of first born, etc. These are all examples of a model I, one-way ANOVA.

In the examples above, we divided data according to one treatment: children belong to different neighborhoods, subjects belong to different ethnicities. In a **two-way ANOVA** the data are divided according to two treatments: children could be divided according to gender and neighborhood, and individuals according to ethnicity and socio-economic status. Because this kind of design requires a relatively large number of subjects which fall into each category (or cell), and because it has a number of assumptions which are difficult to meet in the social sciences, it is infrequently used in anthropology. Therefore, it is not covered here. The difference between a one-way and a two-way **model I ANOVA** is that the first has one, and the second has two treatment effects.

A **model II ANOVA** can be applied to a one- or two-way design. However, in a model II ANOVA there is no treatment effect. Instead, the researcher works with samples that have not been subjected to a treatment but have been obtained from different sources out of the researcher's control. If those sources have affected the subjects, then subjects within each source would be more alike than those of different sources. The purpose of a model II ANOVA is to determine if there is a component of variance due to the source of the subjects.

An example will clarify this. If an archaeologist is working with metric measurements of bi-valve remains (e.g. clams) in an archaeological site which has several stratigraphic levels, the researcher may wonder if bi-valves within a level are less variable than they are among levels. The interest here is simply to know if there is an added component of variation which is due to stratigraphic level. A model II ANOVA would be able to quantify the proportion of variation which is due to stratigraphic level. For the most part the computations of both models do not differ. The only difference is that, if the null hypothesis is rejected, in a model II the researcher proceeds to quantify how much of the total variation is due to the source. This is referred to as computing the added component of variation. Generally in anthropology, there is an interest in finding out if a treatment affects the subjects. Therefore, a model I is usually applied, and is the only model discussed in detail herein.

7.2 ANOVA procedure and nomenclature

Although the ANOVA nomenclature varies among textbooks, it is fairly consistent and easily followed. In this book, we will test the null hypothesis H0: $\mu_1 = \mu_2 = \cdots = \mu_a$ with a **samples** (k is frequently used instead of a). Each sample size will be referred to as n_i, with i having values from 1 to a. The total number of subjects is referred to as Σn (some texts use N instead, although this is misleading since N is the population size). As we learned in the previous chapter, the degrees of freedom (df) are computed as $n_i - 1$. To refer to the sum of variables in each sample, we will use ΣY_i (where the subscript refers to the group), and to refer to the sum of variables across all samples, we will use $\Sigma\Sigma Y$. In the same manner, if we need the sum of the squared variates per sample, we will refer to it as $\Sigma Y^2{}_i$, and we will use $\Sigma\Sigma Y^2$ to refer to the sum of squared observations across samples. For each sample, we can compute the sum of squares

$$SS_i = \Sigma Y^2{}_i - \frac{(\Sigma Y_i)^2}{n_i},$$

which for convenience will be referred to as SS_i. As always, if the sum of squares is divided by the degrees of freedom, a variance is obtained

$$s_i^2 = \frac{\Sigma Y_i^2 - \dfrac{(\Sigma Y_i)^2}{n_i}}{n_i - 1}.$$

In ANOVA nomenclature, however, variances are referred to as mean squares (MS), for reasons explained below.

In a model I ANOVA the null hypothesis states that all samples were obtained from the same population. Thus, the first step in analysis of variance is to estimate the total amount of variation using all subjects without considering group membership. Therefore, we estimate a total sum of squares (SStotal). We then proceed to partition this sum of squares into two components: the component that quantifies the amount of variation within groups (SSerror or SSwithin), and the component that quantifies variation among groups (SSamong). By definition: SStotal = SSwithin + SSamong. Thus, the sums of squares are additive. Basically, the SSamong quantifies the dispersion of sample means around the mean computed using all observations (the grand mean). The SSwithin quantifies the amount of variation found within samples, which can result from the normal variation found in any sample. The reason the SSwithin is also referred to as SSerror is that the best estimate we can make of the expected value of any observation within a sample is the sample mean. The variation around this sample mean can be seen as error, not in the sense of a mistake, but in the sense that data vary.

After the three sums of squares are computed (SStotal, SSwithin, and SSamong), we can compute variances by dividing the sums of squares by degrees of freedom. The reason variance is not called variance in ANOVA (at least in a model I), is that the samples are not randomly obtained from the population proposed by H0. On the contrary, the groups have been subjected to a planned treatment. Thus, we refer to the quotient of the SS divided by the degrees of freedom as mean squares (virtually the mean squared deviation, because we divide by df instead of by n).

For the purposes of ANOVA, we only need to compute two mean squares: the within or error (MSwithin or MSerror), and the among (MSamong). If the null hypothesis is correct, there is no treatment effect, and all samples were obtained from the same population. If that is the case, then subjects within

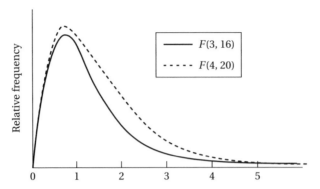

Figure 7.1. The F distribution for several degrees of freedom.

groups are not more alike each other than they are subjects from the other samples. That is, if the null hypothesis is true, there is as much variation within groups as there is among groups. If, on the other hand, the null hypothesis is to be rejected, then the subjects of at least one group are more alike each other than they are to subjects from other groups. If subjects from at least one group are alike and statistically different from subjects of other groups, then the mean of that group should be statistically significantly different from the H0's μ.

We test the H0 by dividing MSamong by MSwithin. Such division yields the F ratio, which is the statistic we use to decide if we reject or accept the H0. The following three situations can occur.

1. If there is more variation within than among samples then we are dividing a MSamong by a larger MSwithin. The product in this case will be an F ratio less than 1, which is *always* non-significant. We would then accept the null hypothesis. Notice that the smallest possible value the F ratio can take is that of 0: mean squares, which quantify variation, can not possibly be negative, and the quotient of two positive numbers can not be negative. This explains why the distribution of the F ratio is cut off at the value of 0 and is not symmetrical (see figure 7.1).

2. If the MSamong and MSwithin are very similar in value then we divide a number by a very similar number. In that case, the product will be an F ratio very close to 1, and will not reject H0.

3. If the MSamong is greater in value than the MSwithin (that is, there is less variation within groups than among groups because the treatment effect affected at least one sample) then we are dividing a large number by a small one. The resulting F ratio will be large, and will fall in the distribution's rejection area, that is, it will be outside 95% of the distribution. As in the t distribution, the distribution of the F ratio depends on the degrees of freedom. However, we need two degrees of freedom to check the table of critical values, since we are working with two variances. Table 3 in Appendix C lists on the top the degrees of freedom associated with the MSamong (the numer-

Table 7.1. *Information presented in an ANOVA table*

Source of variation	df	Sums of squares (SS)	Mean squares (MS)	F	p
Among groups	$a-1$	SSamong	$\dfrac{\text{SSamong}}{a-1}$	$\dfrac{\text{MSamong}}{\text{MSwithin}}$	
Within groups	$\Sigma n-a$	SSwithin	$\dfrac{\text{SSwithin}}{\Sigma n-a}$		
Total variation	$\Sigma n-1$	SSamong + SSwithin			

ator), and on the left side the degrees of freedom associated with the MSwithin (the denominator). If our F ratio is greater than or equal to the critical value, we reject the null hypothesis. It is customary to display all the information used to test the H0 in an ANOVA table (see table 7.1).

The formulae we will use in this book can be used if the samples compared are of the same or different sizes and are more cumbersome than the formulae used for equal sample sizes. However, as presented in the table below, they can be readily computed from the data. All formulae are presented together instead of separately because they form a unit: the ANOVA table.

Formulae 7.1 Formulae for the computation of ANOVA

Source of variation	df	Sums of squares (SS)	Mean squares (MS)	F	p
Among groups	$a-1$	$\displaystyle\sum\dfrac{(\Sigma Y_i)^2}{n_i} - \dfrac{(\Sigma\Sigma Y)^2}{\Sigma n}$	$\dfrac{\text{SSamong}}{a-1}$	$\dfrac{\text{MSamong}}{\text{MSwithin}}$	
Within groups	$\Sigma n-a$	SStotal − SSamong	$\dfrac{\text{SSwithin}}{\Sigma n-a}$		
Total	$\Sigma n-1$	$\Sigma\Sigma Y^2 - \dfrac{(\Sigma\Sigma Y)^2}{\Sigma n}$			

We now do an ANOVA with the data set shown in table 7.2. A cultural anthropologist is interested in studying dietary habits in a community divided into three ethnic groups. The researcher wishes to determine if the groups consume different quantities of rice, a prized food source in two of the groups. The anthropologist records the number of cups of uncooked rice prepared and consumed per week in random samples from the three groups. The households have been matched for the number of persons in the household. The null hypothesis is H0: $\mu_1 = \mu_2 = \mu_3$.

From the data in table 7.2, we have:

$$\Sigma Y_1 = 44 \quad \Sigma Y_2 = 129 \quad \Sigma Y_3 = 145$$

Thus, $\Sigma\Sigma Y = 44 + 129 + 145 = 318$.

Table 7.2 *Number of cups of rice consumed per week*
in a sample of households in three communities

Group A	Group B	Group C
1	7	10
1	8	11
3	7	9
4	10	12
6	8	8
2	9	9
1	7	10
2	9	11
3	9	9
3	10	10
4	6	9
5	9	8
6	11	9
1	10	10
2	9	10
$n_1 = 15$	$n_2 = 15$	$n_3 = 15$
Thus, $\Sigma n = 15 + 15 + 15 = 45$		

$\Sigma Y_1{}^2 = 172 \quad \Sigma Y_2{}^2 = 1{,}137 \quad \Sigma Y_2{}^2 = 1{,}419$

Thus, $\Sigma\Sigma Y^2 = 172 + 1{,}137 + 1{,}419 = 2{,}728.$

$\overline{Y}_1 = 3.93 \quad \overline{Y}_2 = 8.6 \quad \overline{Y}_3 = 9.67$

$s_1 = 1.75 \quad s_2 = 1.40 \quad s_3 = 1.11$

$s_1{}^2 = 3.07 \quad s_2{}^2 = 1.97 \quad s_3{}^2 = 1.23$

It is best to compute first the SStotal, then the SSamong, and last the SSwithin by subtracting the SSamong from the SStotal. Notice, however, that the second term of the SStotal equation is used again in the SSamong. Thus, the latter's formula is not presented in as much detail as is the former.

SStotal:

$$\Sigma\Sigma Y^2 - \frac{(\Sigma\Sigma Y)^2}{\Sigma n} = 2{,}728 - \frac{(318)^2}{45} = 2{,}728 - \frac{101{,}124}{45} = 2{,}728 - 2{,}247.2 = 480.8$$

SSamong:

$$\Sigma\frac{(\Sigma Y_i)^2}{n_i} - \frac{(\Sigma\Sigma Y)^2}{\Sigma n} = \frac{(44)^2}{15} + \frac{(129)^2}{15} + \frac{(145)^2}{15} - 2{,}247.2$$

$$= \frac{1{,}936}{15} + \frac{16{,}641}{15} + \frac{21{,}025}{15} - 2{,}247.2 = 129.1 + 1{,}109.4$$

$$+ 1{,}401.67 - 2{,}247.2 = 2{,}640.13 - 2{,}247.2 = 392.93.$$

SSwithin:

SStotal − SSamong = 480.8 − 392.93 = 87.87

MSamong:

$$\frac{SSamong}{a-1} = \frac{392.93}{2} = 196.47$$

MSwithin:

$$\frac{SSwithin}{\Sigma n - a} = \frac{87.87}{42} = 2.09$$

Fratio:

$$\frac{MSamong}{MSwithin} = \frac{196.47}{2.09} = 94 \text{ with df} = a - 1 \text{ and } \Sigma n - a.$$

Thus df = 2 and 42. For these degrees of freedom the critical value is not listed. Therefore, we use the more conservative values for df = 2, 40. Our F ratio far exceeds the critical value at $\alpha = 0.001$ (cv = 8.25). We therefore reject the null hypothesis: at least one of the samples is different from the other. We display our results in the usual ANOVA table:

Source of variation	df	Sums of squares (SS)	Mean squares (MS)	F	p
Among groups	2	392.93	196.47	94.00	<0.001
Within groups	42	87.87	2.09		
Total	44	480.8			

Practice problem 7.1

An archaeologist is interested in comparing the size of ears of corn found at three contemporaneous archaeological sites located in close proximity. The researcher suspects that one of the communities practiced agriculture more intensely, and could have produced larger ears of corn. In each of the four sites, the following measurements (in centimeters) were taken:

Site A	Site B	Site C	Site D
10	8	10	11
12	10	13	7
8	10	7	10
9	11	7	8
7	8	14	7
13	9	10	10
15	6	9	8
9	11	11	13
10	11	14	8

Site A	Site B	Site C	Site D
11	13	7	8
9	11	9	10
11	10	9	7
8	7	13	10
14	13	11	
10	11	12	
15	16		
13	10		
8	11		
7	9		

$n_1 = 19$ $n_2 = 19$ $n_3 = 15$ $n_4 = 13$

Thus, $\Sigma n = 66$.

$\Sigma Y_1 = 199$ $\Sigma Y_2 = 195$ $\Sigma Y_3 = 156$ $\Sigma Y_4 = 117$

Thus, $\Sigma\Sigma Y = 667$.

$\Sigma Y_1^2 = 2{,}203$ $\Sigma Y_2^2 = 2{,}095$ $\Sigma Y_3^2 = 1{,}706$ $\Sigma Y_4^2 = 1{,}093$

Thus, $\Sigma\Sigma Y^2 = 7{,}097$.

$\bar{Y}_1 = 10.47$ $\bar{Y}_2 = 10.26$ $\bar{Y}_3 = 10.4$ $\bar{Y}_4 = 9$

$s_1 = 2.57$ $s_2 = 2.28$ $s_3 = 2.44$ $s_4 = 1.82$

$s_1^2 = 66$ $s_2^2 = 5.20$ $s_3^2 = 5.97$ $s_4^2 = 3.33$

We follow the usual steps:

SStotal:

$$\Sigma\Sigma Y^2 - \frac{(\Sigma\Sigma Y)^2}{\Sigma n} = 7{,}097 - \frac{(667)^2}{66} = 7{,}097 - \frac{444{,}889}{66} = 7{,}097 - 6{,}740.74 = 356.26$$

SSamong:

$$\Sigma\frac{(\Sigma Y_i)^2}{n_i} - \frac{(\Sigma\Sigma Y)^2}{\Sigma n} = \frac{(199)^2}{19} + \frac{(195)^2}{19} + \frac{(156)^2}{15} + \frac{(117)^2}{13} - 6{,}740.74$$

$$= \frac{39{,}601}{19} + \frac{38{,}025}{19} + \frac{24{,}336}{15} + \frac{13{,}689}{13} - 6{,}740.74$$

$$= 2{,}084.26 + 2{,}001.31 + 1{,}622.4 + 1{,}053 - 6{,}740.74$$

$$= 6{,}760.98 - 6{,}740.74 = 20.24$$

SSwithin:

SStotal − SSamong = 356.26 − 20.24 = 336.02

MSamong:

$$\frac{\text{SSamong}}{a-1} = \frac{20.24}{3} = 6.75$$

MSwithin:

$$\frac{\text{SSwithin}}{\Sigma n - a} = \frac{336.02}{62} = 5.42$$

F ratio:

$$\frac{\text{MSamong}}{\text{MSwithin}} = \frac{6.75}{5.42} = 1.24 \text{ with df} = a - 1 \text{ and } \Sigma n - a$$

Thus df = 3 and 62. For these degrees of freedom the critical value is not listed. Therefore, we use the more conservative values for df = 3, 60. Our F ratio is smaller than the critical value (2.76 at $\alpha = 0.05$). Therefore, we do not reject the null hypothesis: there is no evidence that the size of ears of corn differed among the four sites. We display our results in the usual ANOVA table:

Source of variation	df	Sums of squares (SS)	Mean squares (MS)	F	p
Among groups	3	20.24	6.75	1.24	>0.05
Within groups	62	336.02	5.42		
Total	65	356.26			

7.3 ANOVA assumptions

As should be done with a t test, the assumptions of ANOVA should be tested before the actual test is done. The assumptions of ANOVA remain the same as those mentioned in the previous chapter. Likewise, we only need to test for two assumptions.

   1. *Random sampling.* The samples were collected giving each member of the population an equal chance of being selected.

   2. *Independence of variates.* The collection of an observation does not influence which other observations are collected within and outside of the sample.

   3. *Normality testing.* This is best done with a computer package. SAS performs the normality test with PROC UNIVARIATE.

   4. *Homoscedasticity.* SAS/ASSIST does not provide a menu-driven test of homogeneity of variances in conjunction with the ANOVA menu. Therefore, the Fmax test should be performed by hand, dividing the largest variance by the smallest variance.

   The tests of the normality and homoscedasticity assumptions is illustrated below, in the ANOVA with SAS/ASSIST section.

7.4 Post ANOVA comparison of means

If a model I ANOVA null hypothesis is rejected, we conclude that at least one of the samples did not come from the population proposed by H0. We then need to employ other tests to determine which samples differ. It is not possible to perform t tests repeatedly among the several possible pairs of means because the probability associated with committing a type I error in such a test is one out of 20 ($1/20 = 0.05$). If this is the risk of committing a type I error we are willing to take, then that risk should remain stable *throughout the experiment* or analysis. If we first use a 0.05 α level for the ANOVA, and proceed to compute a number of t tests, we increase our type I error probability for the entire experiment (indeed, if the results of these t tests are plotted, they form a normal distribution, so that 5% of them would be significant as a result of the level we have chosen). Therefore, what we need are tests whose α level is more conservative than 0.05 so that, if we reject the null hypothesis, our chance of having committed an error is still 0.05, not higher.

The tests we will cover in this book are known as post-hoc, un-planned or a-posteriori tests, because they compare all groups after an ANOVA is found to be significant. There are other types of tests known as a-priori or planned tests which are planned before the ANOVA takes place. These would be used in a laboratory setting, in which, for example, a group of animals receives a 'regular' diet (the control), and several groups receive different experimental diets. If the researcher is interested in comparing the effect of the control against *all* experimental groups, then she could perform a planned comparison after the ANOVA. This experimental setting, however, is rarely achieved in the social sciences, in which we would be more interested in comparing all groups should the ANOVA be significant. Therefore, only post-hoc tests will be covered here. There are a large number of tests which can be performed, and SAS provides ready access (under PROC ANOVA, SAS performs, if requested the following tests: Dunnett's, Gabriel's, Ryan–Einot–Gabriel–Welsch, Scheffe, Sidak, GT2, Student–Newman–Keuls, Least-significant-difference, Tukey, and Waller–Duncan). Therefore, it is unnecessary to cover the mathematical computations of more than one such test (for an excellent treatment of multiple comparisons, see Klockars and Sax, 1986).

7.4.1 The Scheffe test

This test was chosen over the many others because it is known to be the post-hoc test which is the most conservative. That is, it is the test which is least likely to produce type I errors. The test proceeds as follows: for each comparison between two group means, a **SSbetween** is computed, and divided by the dfamong to obtain a MSbetween. An F ratio is obtained by dividing the MSbetween by the MSwithin. The F ratio tests the null hypothesis that the two samples were obtained from the same population, and has df $= a - 1$ and $\Sigma n - a$. Only the SSbetween needs to be computed for comparison. Everything else (the MSwithin and the degrees of freedom) is taken from the ANOVA table which has been already computed. The formula for the SSbetween is:

Formula 7.2 Formula for the SSbetween for the Scheffe test

$$\text{SSbetween} = \Sigma \frac{(\Sigma Y_i)^2}{n_i} - \frac{(\Sigma\Sigma Y)^2}{\Sigma n}$$

where only the data of the two groups are considered.

The data first discussed (cups of rice used by three different communities) is used to illustrate the Scheffe test. The following information is necessary:

$n_1 = 15$	$n_2 = 15$	$n_3 = 15$
$\Sigma Y_1 = 44$	$\Sigma Y_2 = 129$	$\Sigma Y_3 = 145$
$\Sigma Y_1^2 = 172$	$\Sigma Y_2^2 = 1,137$	$\Sigma Y_3^2 = 1,419$
$\overline{Y}_1 = 3.93$	$\overline{Y}_2 = 8.6$	$\overline{Y}_3 = 9.67$
$s_1 = 1.75$	$s_2 = 1.40$	$s_3 = 1.11$
$s_1^2 = 3.07$	$s_2^2 = 1.97$	$s_3^2 = 1.23$

The F ratio is compared with the same critical value used previously, since the degrees of freedom are the same: df = 2, 40. All SSbetween will be divided by 2, the dfamong, and all MSbetween by 2.09, the MSwithin.

Comparison of groups 1 and 2. Thus: $\Sigma\Sigma Y = 44 + 129 = 173$, and $\Sigma n = 15 + 15 = 30$.

$$\text{SSbetween} = \left[\frac{(44)^2}{15} + \frac{(129)^2}{15}\right] - \frac{(173)^2}{30} = \left[\frac{1,936}{15} + \frac{16,641}{15}\right] - \frac{29,929}{130}$$

$$= [129.07 + 1,109.4] - 997.63 = 1,238.47 - 997.63 = 240.84$$

$$\text{MSbetween} = \frac{240.84}{2} = 120.42$$

$$F = \frac{\text{MSbetween}}{\text{MSwithin}} = \frac{120.42}{2.09} = 57.62$$

The F ratio far exceeds the critical value at $\alpha = 0.001$ (cv = 8.25). Thus, the null hypothesis that groups 1 and 2 were obtained from the same population is rejected.

Comparison of groups 1 and 3. Thus: $\Sigma\Sigma Y = 44 + 145 = 189$, and $\Sigma n = 15 + 15 = 30$.

$$\text{SSbetween} = \left[\frac{(44)^2}{15} + \frac{(145)^2}{15}\right] - \frac{(189)^2}{30} = \left[\frac{1,936}{15} + \frac{21,025}{15}\right] - \frac{35,721}{30}$$

$$= [129.07 + 1,401.67] - 1,190.7 = 1,530.73 - 1,190.7 = 340.03$$

$$\text{MSbetween} = \frac{340.03}{2} = 170.015$$

$$F = \frac{170.015}{2.09} = 81.35$$

Once again, the null hypothesis is rejected with great confidence: groups 1 and 3 were not obtained from the same population.

Comparison of groups 2 and 3. Thus: $\Sigma\Sigma Y = 129 + 145 = 274$, and $\Sigma n = 15 + 15 = 30$.

$$\text{SSbetween} = \left[\frac{(129)^2}{15} + \frac{(145)^2}{15}\right] - \frac{(274)^2}{30} = \left[\frac{16,641}{15} + \frac{21,025}{15}\right] - \frac{75,076}{30}$$

$$= [1,109.4 + 1,401.67] - 2,502.53 = 2,511.07 - 2,502.53 = 8.53$$

$$\text{MSbetween} = \frac{8.53}{2} = 4.27$$

$$F = \frac{4.27}{2.09} = 2.04$$

The F ratio is smaller than the critical value at $\alpha = 0.05$ (cv $= 3.23$). Thus, the null hypothesis is not rejected. The general conclusion of this study is that group 1 eats a significantly different number of cups of rice per week than groups 2 and 3. The latter two groups do not differ in their consumption.

ANOVA using SAS/ASSIST

The data first discussed (cups of rice used by three different communities) is used. The data were entered into two columns, the first has the number of cups in a household, the second the group membership. There are then $\Sigma n = 45$ lines in the data set. To test the assumptions of normality and of homoscedasticity, a PROC UNIVARIATE (with the normal option) was run in the program manager. The code was submitted, and SAS was referred to the data set which had already been entered. The code is:

```
options linesize = 80 pagesize = 54 date number pageno = 1;
title;
footnote;
proc univariate normal data = BOOK.ANOVA;
  by GROUP;
run;
quit;
```

For the sake of space, part of the PROC UNIVARIATE output is not reproduced here. The output is:

```
--------------community to which the household belongs = 1--------------
                         Univariate Procedure
Variable = CUPS            number of cups of rice per week
                                Moments

          N                15   Sum Wgts          15
          Mean       2.933333   Sum               44
          Std Dev     1.75119   Variance    3.066667 ←for Fmax
          Skewness   0.576906   Kurtosis   -0.77931
          USS             172   CSS         42.93333
          CV         59.69966   Std Mean    0.452155
          T:Mean = 0 6.487446   Pr>|T|        0.0001
```

```
--------------community to which the household belongs = 1--------------
                          Univariate Procedure
    Variable = CUPS        number of cups of rice per week
                                 Moments
                Num ^ = 0        15   Num > 0          15
                M(Sign)         7.5   Pr > = |M|    0.0001
                Sgn Rank         60   Pr > = |S|    0.0001
                W:Normal  0.895134    Pr < W        0.0810 ←normal(>0.05)


--------------community to which the household belongs = 2--------------
                                 Moments
                N                15   Sum Wgts         15
                Mean            8.6   Sum             129
                Std Dev    1.404076   Variance   1.971429
                Skewness  -0.23582    Kurtosis  -0.64636
                USS            1137   CSS            27.6
                CV         16.32646   Std Mean   0.362531
                T:Mean = 0 23.72212   Pr > |T|      0.0001
                Num ^ = 0        15   Num > 0          15
                M(Sign)         7.5   Pr > = |M|    0.0001
                Sgn Rank         60   Pr > = |S|    0.0001
                W:Normal  0.944154    Pr < W        0.4236 ←normal(>0.05)


--------------community to which the household belongs = 3--------------
                                 Moments
                N                15   Sum Wgts         15
                Mean       9.666667   Sum             145
                Std Dev    1.112697   Variance   1.238095 ←for Fmax
                Skewness   0.412134   Kurtosis  -0.00956
                USS            1419   CSS        17.33333
                CV         11.51066   Std Mean   0.287297
                T:Mean = 0 33.64692   Pr > |T|      0.0001
                Num ^ = 0        15   Num > 0          15
                M(Sign)         7.5   Pr > = |M|    0.0001
                Sgn Rank         60   Pr > = |S|    0.0001
                W:Normal   0.92555    Pr < W        0.2319 ←normal(>0.05)
```

The test indicates that the data are normal (although the first sample has a W statistic that approaches the critical value). To perform the Fmax test, the larger variance is divided by the smaller variance, and the F ratio compared to the Fmax table with df $= 15–1 = 14$ and $a = 3$. We use the closest degrees of freedom listed in the table, namely 12 and 3. At these degrees of freedom, the critical value is 4.16. Thus, Fmax $= 3.066667 / 1.238095 = 2.48$. The variances are homogeneous.

To perform an ANOVA, the user follows this path: data analysis, ANOVA, ANOVA. At the ANOVA window, the user specifies the dependent variable (cups) and the classificatory variable (group). Unfortunately, the menu-driven commands do not include the possibility of obtaining post-hoc tests. To obtain them, the SAS-generated code is transformed by adding a means statement. The code is:

```
options linesize = 80 pagesize = 54 date number pageno = 1;
title;
footnote;
proc anova data = BOOK.ANOVA;
  class GROUP;
```

```
  model CUPS = GROUP ;
  means group /scheffe; ←statement added to original code
run;
quit;
```

Below is the output, including the Scheffe test:

```
                Analysis of Variance Procedure
                  Class Level Information

           Class    Levels   Values
           GROUP        3    1 2 3

        Number of observations in data set = 45
```

```
Analysis of Variance Procedure
Dependent Variable: CUPS number of cups of rice per week
                       Sum of           Mean
Source          DF    Squares         Square  F Value  Pr>F

Model            2  392.93333333  196.46666667   93.91  0.0001

Error           42   87.86666667    2.09206349

Corrected Total 44  480.80000000
```

```
                Analysis of Variance Procedure
                Scheffe's test for variable: CUPS
```

NOTE: This test controls the type I experimentwise error rate but
      generally has a higher type II error rate than REGWF for all
      pairwise comparisons

```
         Alpha = 0.05 df = 42 MSE = 2.092063
            Critical Value of F = 3.21994
         Minimum Significant Difference = 1.3403
```

Means with the same letter are not significantly different.

```
   Scheffe Grouping    Mean    N  GROUP

            A    9.6667   15   3←⎤
            A                      Not different
            A    8.6000   15   2←⎦
            B    2.9333   15   1
```

7.5 A research example of an ANOVA

In their paper entitled '"Race" specificity and the femur/stature ratio',
Feldesman and Fountain (1996) investigate whether differences in the
femur/stature ratio in three quasi-geographic 'races' are statistically
significant. Using a large data set consisting of 55 sample means of modern
human populations assembled into three groups (thus, $a = 3$ and $\Sigma Y = 55$),
Feldesman and Fountain test the null hypothesis that the groups were
obtained from the same population. They followed the ANOVA with the
Tukey multiple comparisons test, which identified which groups differed
significantly. When admixed populations (US Mexicans and Puerto Ricans)

were excluded, the null hypothesis is rejected with a probability of $p = 0.0003$ (another comparison had included these groups classified as 'Asian'). The Tukey post-hoc test finds significant differences between 'Blacks' and 'Whites' ($p = 0.012$), and between 'Blacks' and 'Asians' ($p < 0.0005$), while the 'White'–'Asian' difference is not significant ($p = 0.09$) (Feldesman and Fountain, 1996: 216). In the ANOVA in which the admixed groups had been included as Asians, the comparison between 'Asians' and 'Whites' was not significant either, although even less. The authors mention in their discussion that the ANOVA analysis confirms that there are statistically significant differences in the mean femur/stature ratios of the three geographic 'races' constructed for this analysis. At the same time, multiple comparison tests, done to resolve the significant F tests from the ANOVAs, make it clear that the 'Black' femur/stature ratio is the outlier; neither 'White' nor 'Asian' ratios are statistically distinct from one another (Feldesman and Fountain, 1996: 218–219).

7.6 Key concepts

One-way ANOVA
Model I and II ANOVA
Why is the F ratio not symmetrical?
Assumptions of ANOVA
Post-hoc tests, the Scheffe test

7.7 Exercises

1. The following data are the weights in grams of Rangia shells from three levels from Depot Creek site (White, work in progress). Perform an ANOVA, and test its assumptions.

Weight in grams	Stratigraphic level	Weight in grams	Stratigraphic level	Weight in grams	Stratigraphic level
6.5	1	9.0	3	5.0	4
6.7	1	17.2	3	3.1	4
8.9	1	4.4	3	3.1	4
10.8	1	6.7	3	3.2	4
5.0	1	10.7	3	3.0	4
25.0	1	5.5	3	3.8	4
12.0	1	19.4	3	3.3	4
12.6	1	12.0	3	9.4	4
6.3	1	7.6	3	4.3	4
5.8	1	3.5	3	1.5	4
6.8	1	4.8	3	4.1	4

Weight in grams	Stratigraphic level	Weight in grams	Stratigraphic level	Weight in grams	Stratigraphic level
5.0	1	3.4	3	6.0	4
8.0	1	4.2	3	5.1	4
8.1	1	3.3	3	7.2	4
12.0	1	3.5	3	10.8	4
8.5	1	13.3	3	2.1	4
10.8	1	13.3	3	10.7	4
4.2	1	12.0	3	6.3	4
12.0	1	8.5	3	10.1	4
6.4	1	7.2	3	21.2	4
4.9	1	14.4	3	23.8	4
7.0	1	7.1	3	13.0	4
4.4	1	10.7	3	13.6	4
6.2	1	9.0	3	18.0	4
3.7	1	14.3	3	10.9	4
5.1	1	5.8	3	13.2	4
5.4	1	13.3	3	14.7	4
6.2	1	14.7	3	12.1	4
6.1	1	9.3	3	11.5	4
4.2	1	16.4	3	23.1	4
5.9	1	5.7	3	13.5	4
7.6	1	5.3	3	12.5	4
1.9	1	9.3	3	12.1	4
2.3	1	5.4	3	11.5	4
2.6	1	10.2	3	23.1	4
5.3	1	6.2	3	12.0	4

$$\Sigma n_i = 108$$

2. A researcher is interested in determining if the mean birth weight in three ethnic groups residing in the same area, and attending the same public hospital, differ significantly. The birth weight (rounded to the nearest pound) of a random sample of healthy babies from the three groups is shown below. Perform an ANOVA, and test its assumptions.

Group 1	Group 2	Group 3
6	8	5
7	5	6
8	7	6
6	6	7
7	8	5
7	8	6
8	6	6
7	7	7
9	7	5
7	6	6
8	8	5

Group 1	Group 2	Group 3
7	6	5
8	7	6
7	7	5
8	7	5
6	8	6
7	7	6

8 | Non-parametric comparison of samples

Researchers should always test the assumptions of parametric tests. What should be done if the assumptions are violated? The data can be transformed, or a non-parametric test applied to the raw data. The reader is advised that if the data set with which he is working is fairly large, then he should transform it, fix the assumption violations, and apply a parametric test. With a computer, this can be done fairly efficiently, and will take less time than applying a non-parametric test, whose hand calculations can be cumbersome for large data sets. If the data set is small, however, the researcher may be better off applying a non-parametric test.

Non-parametric tests are distribution-free. That is, they do not make any assumptions about the distribution of the population from which the samples were collected. Hence, these tests can be applied to non-normal data. Non-parametric tests are so-called because their null hypothesis does not specify a value for the population's parameter. Hence, the null hypothesis will not be written in parametric notation. Rather, it will simply be stated as, for example, there is no treatment effect, or the samples were obtained from the same population, etc.

Non-parametric tests are particularly well suited for many social science situations in which the sample sizes are exceedingly small: a paleo-anthropologist may have small samples of fossils, or an archaeologist may wish to compare a few projectile points from several sites. If a researcher is working with such small samples, then she may as well not even attempt to apply a parametric test, but simply choose a non-parametric one. A further advantage of non-parametric tests is that they are easily computed by hand. This is why, if a researcher has a small, non-normal data set, he is well advised to perform a non-parametric test instead of trying to transform the data in order to apply a parametric one. It would probably be faster to do the non-parametric test by hand than transforming the data and applying a parametric test in the computer.

Another advantage of non-parametric tests is that they do not require data to be measured precisely. Since the data only need to be able to be ranked, they can be variables such as prestige in a community, a culture's commitment to animal domestication, etc. In other words, non-parametric tests can be applied to many sets of data collected on culture, ideology, arts, etc. Also, these tests should be preferred when comparing derived variables, for example rates, which are the quotient of a division as opposed to an actual measure of an object.

For all its advantages, non-parametric tests are known to be not as statistically powerful as parametric tests are. Therefore, the latter should be preferred over the former if the data do not violate the assumptions, and if the sample sizes are not too small.

This chapter covers three non-parametric tests, to be used instead of an un-paired t test, a one-way model I ANOVA, and a paired t test. These tests have a procedure in common: they rank the data before the statistics are computed. Thus, we first learn how to rank data.

8.1 Ranking data

The procedure for ranking data in these tests is basically the same, except that two of the tests to be covered here consider the sign of the digit, whereas the other one works with the absolute value of the numeral. Let us practice both kinds of ranking with the following numbers: $-2, 6, 1$. If we rank considering the sign, then the ranking should yield:

Number	Rank
-2	1
1	2
6	3

That is, if we consider the sign, then -2 is less than 1. However, if we do not consider the sign, we need to rank the data according to their absolute value. Therefore, our ranking results should be:

Number	Rank
1	1
-2	2
6	3

For the time being, we will practice ranking without regard to the sign of the data. A commonly confronted situation when ranking is the presence of ties. When observations are tied, the average of their rank is assigned to all of them. Therefore, there is an initial step: we first assign ranks as if the variates were not tied, then we average these ties. For example, given the following data, we take these steps.

1. We first order the variates from lowest to highest (using the absolute value).
2. We then assign the initial ranks.
3. In case of ties, we compute the average of the initial ranks, and assign them as final ranks.

Raw data	Ordered data	Initial rank	Average ranks of tied variates	Final rank
3	0	1		1
4	1	2		2
9	3	3		3.5
0	3	4	$\left.\dfrac{3+4}{2}\right\} = 3.5$	3.5
1	4	5		5
3	6	6		6
9	7	7		7
6	9	8		9
7	9	9	$\left.\dfrac{8+9+10}{3}\right\} = 9$	9
9	9	10		9

8.2 The Mann–Whitney U test for an un-matched design

This test can be used instead of an un-paired t test if, for any of the reasons discussed in the introduction, the latter is not appropriate. This test only makes two assumptions.

1. The underlying level of measurement is continuous, although the units of observations are discrete. Therefore, ties in ranking occur because of error of measurement. The researcher's only concern about this assumption is if the data contain too many ties.
2. The samples are independent. Thus, this test is not appropriate for a paired design.

The null hypothesis is that the population from which the samples are obtained is the same. If the null hypothesis is true, and the two samples were ranked together, the observations of both samples would be expected to be equally mixed. If however, the samples came from two different populations, they would not be expected to mix, but to be on opposite sides of the ranked column of data. The test determines how evenly mixed the scores are from both samples. The Mann–Whitney test can be applied to samples of equal or different sizes.

The test proceeds as follows.

1. Rank the observations of both samples together. Ranking is done taking into consideration the sign of the observation.
2. For each sample (1 and 2), compute n_1 and n_2 and the sum of the ranks ΣR_1 for the first group.
3. Compute the U statistic for both groups as follows:

Formula 8.1 Formula for the *U* statistic

$$U_1 = (n_1)(n_2) + \frac{(n_1)(n_1 + 1)}{2} - \Sigma R_1, \quad \text{and } U_2 = (n_1)(n_2) - U_1$$

4. Choose the *lesser* value of *U*, either U_1 or U_2, for significance testing. Using table 5 in Appendix C, compare the obtained *U* with the critical value for the appropriate sample sizes, but **use a different decision rule**: if the test statistic at $\alpha = 0.05$ *is less than or equal to* the critical value, reject the null hypothesis. This table can be used if n_1 and n_2 are less than or equal to 20. For larger sample sizes, the *U* statistic can be transformed into a *t* or *z* score, and a *t* or normal table used (see Sokal and Rohlf, 1981). Such transformation is not discussed here because samples whose sizes are over 20 can be subjected to a test of normality and, if the normality assumption is rejected, the data can be transformed, and a parametric test used. As was mentioned in the introduction, the hand calculations of non-parametric tests are quite simple for small data sets, but not for large ones.

The use of this table presents a major departure from the previous use of statistical tables. The reason for this different approach should be explained with an example, which shows two exteme cases. In the first, the samples have ranks that do not mix at all; and in the second, the samples are ranked totally mixed. We follow the steps outlined above, starting with the first case:

1. Rank the observations of both samples together.

Rank	Group membership
1	1
2	1
3	1
4	1
5	1
6	1
7	1
8	1
9	2
10	2
11	2
12	2
13	2
14	2
15	2

2. For each sample (1 and 2), compute n_1 and n_2, and the sum of the ranks ΣR_1 for the first group. Thus, $n_1 = 8$, $n_2 = 7$ and $\Sigma R_1 = 1 + 2 + 3 + 4 + 5 + 6 + 7 + 8 = 36$.

3. Compute the *U* statistic for both groups:

$$U_1 = (8)(7) + \frac{(8)(8+1)}{2} - 36 = 56 + 36 - 36 = 56,$$

and $U_2 = (8)(7) - 56 = 56 - 56 = 0.$

4. Choose the *lesser* value of U, either U_1 or U_2, for significance testing. The critical value at $\alpha = 0.05$ for a two-tail test where $n_1 = 8$, $n_2 = 7$ is 10. Therefore, the null hypothesis that the two samples were obtained from the same population is rejected.

In general, it is known that $U_1 + U_2 = (n_1)(n_2)$. In this example for instance, $(n_1)(n_2) = (8)(7) = 56$, and $U_1 + U_2 = 56 + 0 = 56$. In an extreme case in which there is no overlap between the two samples, the obtained U used for hypothesis testing is 0.

Now the second example is presented. Here, the two samples are mixed.

1. Rank the observations of both samples together.

Rank	Group membership
1	1
2	2
3	1
4	2
5	1
6	2
7	1
8	2
9	1
10	2
11	1
12	2
13	1
14	2
15	1

2. For each sample (1 and 2),compute n_1 and n_2 and the sum of the ranks ΣR_1 for the first group. Thus, $n_1 = 8$, $n_2 = 7$ and $\Sigma R_1 = 1 + 3 + 5 + 7 + 9 + 11 + 13 + 15 = 64$.

3. Compute the U statistic for both groups:

$$U_1 = (8)(7) + \frac{(8)(8+1)}{2} - 36 = 56 + 36 - 64 = 28,$$

and $U_2 = (8)(7) - 28 = 56 - 28 = 28.$

4. Choose the *lesser* value of U, either U_1 or U_2, for significance testing. The critical value at $\alpha = 0.05$ for a two-tail test where $n_1 = 8$, $n_2 = 7$ is 10. In this case, either U may be used for hypothesis testing since they have the same value.

The null hypothesis is accepted because the test statistic is greater than the critical value.

The reader should note that $(n_1)(n_2) = (8)(7) = 56$ and that and $U_1 + U_2 = 28 + 28 = 56$. In this extreme example in which the variates were mixed by alternating group membership, the two U values were $(n_1)(n_2)/2$. This is the highest possible value the U can take, if the two groups do not differ. This is the reason why the null hypothesis in a Mann–Whitney U test is tested by determining how small the smallest U is. The closer it is to zero, the more likely the hypothesis is to be rejected. The test is now illustrated with a more realistic example.

Assume that an anthropologist is investigating the social role of young children in two ethnically distinct groups in a community. The researcher is interested in investigating if the children in the two groups acquire a clear economic/social responsibility at different ages, or if the two communities assign essentially the same roles to young children. The anthropologist decides to study if children who accompany their mothers in their daily routine are of significantly different ages. These children essentially do not have any clearly designated tasks, outside of being with their mothers and playing with their friends. A problem the anthropologist encounters, however, is that there are no records to validate the children's ages, which must be estimated by the researcher in consultation with the mothers. To minimize error in the estimation, the researcher limits her study to one gender of children only. A non-parametric test should be used in this case because the sample sizes are small and the data are estimates, not true measures of the actual age of the children. Below are the data:

Child's age in years (estimate)	Group
1	1
2	1
2	1
4	1
5	1
6	1
2	2
3	2
4	2
5	2
5	2

The steps outlined above are followed.

1. Rank the observations of both samples together.

Child's age in years (estimate)	Group	Initial rank	Final rank
1	1	1	1
2	1	2	3
2	2	3 $\left.\dfrac{2+3+4}{3}\right\} =$	3
2	1	4	3
3	2	5	5
4	2	6 $\left.\dfrac{6+7}{2}\right\} =$	6.5
4	1	7	6.5
5	2	8	9
5	1	9 $\left.\dfrac{8+9+10}{3}\right\} =$	9
5	2	10	9
6	1	11	11

2. For each sample (1 and 2), compute $n_1 = 6$ and $n_2 = 5$ and the sum of the ranks

$$\Sigma R_1 = 1 + 3 + 3 + 6.5 + 9 + 11 = 33.5, \Sigma R_2 = 3 + 5 + 6.5 + 9 + 9 = 32.5.$$

3. Compute the U statistic for both groups 1 and 2, where

$$U_1 = (6)(5) + \frac{(6)(6+1)}{2} - 33.5 = 17.5 \text{ and } U_2 = (6)(5) - 17.5 = 12.5.$$

4. Choose the lesser value of U, either U_1 or U_2, for significance testing. Thus, we choose U_2, which is not less than or equal to the critical value at $\alpha = 0.05$ (cv for $n_1 = 6$ and $n_2 = 5$ is 3). Therefore, we do not reject the null hypothesis, and conclude that the children who accompany their mothers and have not been assigned a specific economic task have the same age in the two different ethnic groups (the reader should check that $U_1 + U_2 = (n_1)(n_2) = 17.5 + 12.5 = (6)(5) = 30 = 30$).

Practice problem 8.1

The data for this practice problem consist of oyster shell metric measurements from the Van Horn Creek site (data provided by White, work in progress), and are the weight in grams of individual oyster 'bottom' valves. Data were obtained from different stratigraphic levels, and we wish to test if the oysters differ between levels 1 and 2 for their weight. Because of the small sample sizes, we apply a non-parametric test. The data are:

Level 1	Level 2
112.1	58.9
74.1	55.8
68.9	8.4

Level 1	Level 2
35.2	67.5
55.0	93.0
21.2	25.6
61.5	41.8
9.1	
40.9	
26.1	
30.0	

The usual steps are followed.

1. Rank the observations of both samples together. The ranking was done by SAS. The rank is called 'new'.

```
Obs.  Weight  Level  New
 1      8.4     2     1
 2      9.1     1     2
 3     21.2     1     3
 4     25.6     2     4
 5     26.1     1     5
 6     30.0     1     6
 7     35.2     1     7
 8     40.9     1     8
 9     41.8     2     9
10     55.0     1    10
11     55.8     2    11
12     58.9     2    12
13     61.5     1    13
14     67.5     2    14
15     68.9     1    15
16     74.7     1    16
17     93.0     2    17
18    112.1     1    18
```

2. For each sample compute $n_1 = 11$ and $n_2 = 7$ and the sum of the ranks for the first group: $\Sigma R_1 = 108$.

3. Compute the *U* statistic for both groups 1 and 2, where

$$U_1 = (11)(7) + \frac{(11)(11+1)}{2} - 108 = 35, \quad \text{and } U_2 = (7)(11) - 35 = 42.$$

4. Choose the lesser value of *U*, either U_1 or U_2, for significance testing. In this case, we choose $U_1 = 35$ which is compared with the critical value for $n_1 = 11$ and $n_2 = 7$ (cv = 16). We fail to reject the null hypothesis, and conclude that the oysters found at the two stratigraphic levels do not differ significantly in their weight.

....

8.3 A research example of the Mann–Whitney *U* test

The factors that influence a mother's decision to breastfeed, and for how long, are of interest to applied researchers from several disciplines. Quarles et al. (1994) investigated if exposure to the services of a certified lactation consultant (CLC) influenced mothers' duration of breastfeeding. Specifically, they tested the following hypothesis: 'A greater proportion of mothers at Hospital 1 (with CLC) will attain their intended length of breastfeeding, compared to mothers at Hospital 2 (non-CLC).' Intended length was measured by subtracting the actual duration of breastfeeding from the intended duration. Since this is a variable derived from a subtraction, a non-parametric test was appropriate. The research hypothesis is justifiably one-tailed, since there is only interest in demonstrating a positive effect of exposure to a CLC on breastfeeding duration. The researchers found a significant difference in attainment of intended breastfeeding length in mothers who were exposed to the lactation consultant ($z = -1.94$, $p = 0.03$). Apparently, the U statistic was transformed into a z score, a common practice in the literature.

8.4 The Kruskal–Wallis instead of a one-way, model I ANOVA

This test rests on the same principles as does the Mann–Whitney U test, except that it applies to more than two groups: if the null hypothesis that the groups were obtained from the same population is true, then the observations from all groups would be expected to be ranked randomly, as opposed to the observations from each group being ranked together. Again, the assumption of independence among the groups holds, as does the assumption that ties are due to measurement error. Indeed, if there are ties, the computation of the statistic is slightly different, as will be seen below. The test can be applied to samples of equal or different size. Each group's size will be denoted as n_i, where i ranges from 1 to a, where a is the number of groups compared. The sum of all sample sizes will be referred to as $\Sigma\Sigma n$. The test proceeds as follows.

1. Rank the observations of all samples together. Ranking is done taking into consideration the sign of the observation.
2. Compute the sample size and the sum of the ranks ΣR_i for all the groups.
3. Compute the H statistic as follows:

Formula 8.2 The H statistic for the Kruskal–Wallis test (no ties)

$$H = \left[\left(\frac{12}{(\Sigma\Sigma n)(\Sigma\Sigma n + 1)} \right) \left(\Sigma \frac{(\Sigma R_i)^2}{n_i} \right) \right] - 3(\Sigma\Sigma n + 1)$$

If there are no ties in the data, the H statistic is compared with the χ^2 table (table 6 in Appendix C) with degrees of freedom $= a - 1$ (the χ^2 distribution will be discussed later in the book). However, if there are ties in the ranking procedure, than H must be divided by a correction term D. The adjusted H statistic is then compared with the critical value in the χ^2 table with degrees of freedom $= a - 1$. The correction term takes into consideration the *number* of variates which were tied (t_j), where j is the number for groups of tied variates. For example:

$$\left.\begin{array}{l}1\\1\\1\end{array}\right\}t_1=3$$

$$\left.\begin{array}{l}2\\2\end{array}\right\}t_2=2$$

$$3$$

$$4$$

$$\left.\begin{array}{l}5\\5\\5\\5\end{array}\right\}t_3=4$$

There are 11 observations, and there are three groups of tied variates. Thus, $j=3$. In the first group, there were three tied variates. Therefore, $t_1=3$. In the second and third groups there were two and four tied variates respectively, so $t_2=2$ and $t_3=4$. The correction for the H statistic considers the number of tied variates, and the number of groups of tied variates, by computing for each group the function T_j. The complete formula for the correction term is:

....

Formula 8.3 The H statistic for the Kruskal–Wallis test when ties occur

$$H=\frac{H}{D}, \quad \text{where} \quad D=1-\frac{\Sigma T_j}{(\Sigma\Sigma n_i-1)(\Sigma\Sigma n_i)(\Sigma\Sigma n_i+1)}$$

T is computed for each group as $(t_j-1)(t_j)(t_j+1)$, where ΣT_j is the sum of T across all groups of tied variates, and $\Sigma\Sigma n$ is the sum of all sample sizes

....

We now practice the computation of T and ΣT_j for:

$$\left.\begin{array}{l}1\\1\\1\end{array}\right\}t_1=3 \quad T_1=(t_j-1)(t_j)(t_j+1)=(3-1)(3)(3+1)=24$$

$$\left.\begin{array}{l}2\\2\end{array}\right\}t_2=2 \quad T_2=(t_j-1)(t_j)(t_j+1)=(2-1)(2)(2+1)=6$$

$$3$$

$$4$$

$$\left.\begin{array}{l}5\\5\\5\\5\end{array}\right\}t_3=4 \quad T_3=(t_j-1)(t_j)(t_j+1)=(4-1)(4)(4+1)=60$$

For this small data set, then

$$D = 1 - \frac{\Sigma T_j}{(\Sigma\Sigma n_i - 1)(\Sigma\Sigma n_i)(\Sigma\Sigma n_i + 1)}$$

$$= 1 - \frac{24 + 6 + 60}{(11-1)(11)(11+1)} = 1 - \frac{90}{1{,}320} = 1 - 0.068 = 0.932$$

We now practice the computation of the Kruskal–Wallis test with the same study which tested the hypothesis that children from two different ethnic groups who accompany their mothers, and do not have a clearly assigned economic task, are of the same age. However, we now expand our study to include a third ethnic group. The data are:

Age (estimate)	Group
1	1
2	1
2	1
4	1
5	1
6	1
2	2
3	2
4	2
5	2
5	2
1	3
1	3
2	3
3	3
3	3
1	3

We follow the already mentioned steps.

1. Rank the observations of all samples together.

Obs	Age	Group	Rank
1	1	1	2.5
2	1	3	2.5
3	1	3	2.5
4	1	3	2.5
5	2	1	6.5
6	2	1	6.5
7	2	2	6.5
8	2	3	6.5
9	3	2	10.0

Obs	Age	Group	Rank
10	3	3	10.0
11	3	3	10.0
12	4	1	12.5
13	4	2	12.5
14	5	1	15.0
15	5	2	15.0
16	5	2	15.0
17	6	1	17.0

2. Compute the sample size and the sum of the ranks ΣR_i for all the groups.

Group 1   Group 2   Group 3
$n_1 = 6$ $n_2 = 5$ $n_3 = 6$ Therefore: $\Sigma\Sigma n = 17$
$\Sigma R_1 = 60$ $\Sigma R_2 = 59$ $\Sigma R_3 = 34$

3. Compute the H statistic as follows:

$$H = \left[\left(\frac{12}{(17)(17+1)}\right)\left(\frac{(60)^2}{6} + \frac{(59)^2}{5} + \frac{(34)^2}{6}\right)\right] - 3(17+1)$$

$$= \left[\left(\frac{12}{306}\right)(600 + 696.2 + 192.67)\right] - 3(18)$$

$$= [(0.03921)(1,488.87)] = 58.378 - 54 = 4.38$$

This H statistic must be corrected because the data set has ties. There are five groups of tied variates, for each of which we compute T.

$\left.\begin{array}{l}1\\1\\1\\1\\1\end{array}\right\}t_1 = 4 \quad T_1 = (4-1)(4)(4+1) = 60$

$\left.\begin{array}{l}2\\2\\2\\2\end{array}\right\}t_2 = 4 \quad T_2 = (4-1)(4)(4+1) = 60$

$\left.\begin{array}{l}3\\3\\3\end{array}\right\}t_3 = 3 \quad T_3 = (3-1)(3)(3+1) = 24$

$\left.\begin{array}{l}4\\4\end{array}\right\}t_4 = 2 \quad T_4 = (2-1)(2)(2+1) = 6$

$\left.\begin{array}{l}5\\5\\5\end{array}\right\}t_5 = 3 \quad T_5 = (3-1)(3)(3+1) = 24$

6

The correction term is thus computed as follows:

$$D = 1 - \frac{60 + 60 + 24 + 6 = 24}{(17-1)(17)(17+1)} = 1 - \frac{174}{4{,}896} = 1 - 0.035 = 0.96$$

Last, we compute the corrected H, and compare it with the χ^2 table with degrees of freedom $= a - 1$. $H = \dfrac{H}{D} = \dfrac{4.38}{0.96} = 4.5625$. The critical value at $\alpha = 0.05$ for df $= 2$ is 5.991. Therefore, we do not reject the null hypothesis that the three groups were obtained from the same population.

If a Kruskal–Wallis test is performed and the null hypothesis is rejected, the investigator needs to determine which group(s) is different from which. Generally, this test will simply be the Mann–Whitney U test which we just learned. However, if the comparison involves samples of equal size, whose n is at least 8, then another formulation of the Mann–Whitney is more appropriate (see Sokal and Rohlf, 1981).

Practice problem 8.2

We now practice the Kruskal–Wallis test with the archaeological data set collected at the Van Horn Creek site consisting of the weight in grams of the bottom valve of oysters (White, work in progress). We include in our test the two stratigraphic levels tested in the previous section, but add a third one. The data now are:

Weight in grams	Stratigraphic level
112.1	1
74.7	1
68.9	1
35.2	1
55.0	1
21.2	1
61.5	1
9.1	1
40.9	1
26.1	1
30.0	1
58.9	2
55.8	2
8.4	2
67.5	2
93.0	2
25.6	2
41.8	2
45.1	3
78.1	3
18.6	3
22.0	3
36.4	3
89.3	3
19.3	3
39.7	3

```
Weight   Stratigraphic
in grams     level
  21.2         3
  17.1         3
   6.5         3
```

The usual steps are followed.

1. Rank the observations of all samples together.

Obs.	Weight	Level	Rank
1	6.5	3	1.0
2	8.4	2	2.0
3	9.1	1	3.0
4	17.1	3	4.0
5	18.6	3	5.0
6	19.3	3	6.0
7	**21.2**	**1**	**7.5**
8	**21.2**	**3**	**7.5**
9	22.0	3	9.0
10	25.6	2	10.0
11	26.1	1	11.0
12	30.0	1	12.0
13	35.2	1	13.0
14	36.4	3	14.0
15	39.7	3	15.0
16	40.9	1	16.0
17	41.8	2	17.0
18	45.1	3	18.0
19	55.0	1	19.0
20	55.8	2	20.0
21	58.9	2	21.0
22	61.5	1	22.0
23	67.5	2	23.0
24	68.9	1	24.0
25	74.7	1	25.0
26	78.1	3	26.0
27	89.3	3	27.0
28	93.0	2	28.0
29	112.1	1	29.0

2. Compute the sample size and the sum of the ranks ΣR_i for all groups.

Group 1	Group 2	Group 3	
$n_1 = 11$	$n_2 = 7$	$n_3 = 11$	Therefore: $\Sigma\Sigma n = 29$
$\Sigma R_1 = 181.5$	$\Sigma R_2 = 121$	$\Sigma R_3 = 132.5$	

3. Compute the H statistic as follows:

$$H = \left[\left(\frac{12}{(29)(29+1)} \right) \left(\frac{(181.5)^2}{11} + \frac{(121)^2}{7} + \frac{(132.5)^2}{11} \right) \right] - 3(29+1)$$

$$= \left[\frac{12}{870} \right] [2{,}994.75 + 2{,}091.6 + 1{,}596.02] - 90 = (0.01379)(6{,}682.3441) - 90$$

$$= 92.15 - 90 = 2.15$$

Because there was one tie, the H statistic must be corrected by dividing it by D. We first compute t_j for the only group of tied variates, that is, for:

$\left.\begin{array}{l}21.2 \\ 21.2\end{array}\right\}$ $t_1 = 2 = (2-1)(2)(2+1) = 6$. Therefore, T, the sum of t_j, is 6.

$$D = 1\frac{6}{(29-1)(29)(29+1)} = \frac{6}{24,360} = 1 - 0.00024 = 0.9997$$

Last, we compute the corrected H, and compare it with the χ^2 table with degrees of freedom $= a - 1$, or $3 - 1 = 2$.

$$H = \frac{H}{D} = \frac{2.15}{0.9997} = 2.151$$

Since this H statistic is less than the critical value at $\alpha = 0.05$ (cv $= 5.991$), we fail to reject the null hypothesis. The samples of oysters appear to have been obtained from a single population.

8.5 A research example of the Kruskal–Wallis test

It is known that lipoprotein lipase (LPL) affects the level of low-density lipoprotein (LDL), and that elevated LDL cholesterol is strongly associated with an increased risk of coronary artery disease. It is therefore of interest to determine if an association exists between LPL gene polymorphisms and triacylglyceride levels. Mitchell et al. (1994) investigate if two polymorphisms at the LPL locus are correlated with plasma lipoprotein phenotypes in male migrants to Australia from Italy and Greece. Since the distribution of plasma lipid levels are usually not normally distributed, a non-parametric test is desired. The Kruskal–Wallis tests demonstrated a significant association of the *Hind* III polymorphisms with 'both triacylglycerides ($p = 0.03$) and HDL cholesterol ($p = 0.008$) but not with LDL cholesterol or total cholesterol' (Mitchell et al, 1994: 389).

8.6 The Wilcoxon signed-rank test for a paired design

The last non-parametric test covered in this chapter is one to be used instead of a paired t test. The reader should recall that an assumption of a paired t test is that the differences be normally distributed. If this assumption is violated, or if the data set is very small, or the measures are approximate, then a non-parametric test should be used. As in the paired t test, the two-tailed null hypothesis is that the differences are not significantly different from 0. However, the null hypothesis is not expressed in parametric terms. Frequently the H0 will be stated in a one-tail manner, since the researcher has reason to know that the treatment will only affect the subjects in one direction. The test does assume that each individual was randomly selected, and that ties are the result of imperfect measures.

The Wilcoxon signed-rank test computes differences between the two measures taken on the individual (or between the measures taken on the matched individuals, whatever the case may be). Such differences are then ranked *without regard to the sign (although the sign of the difference will be used later).* That is, -1 and 1 would be tied and thus ranked equally.

Two extreme cases may occur: if the treatment affects all subjects to change either positively or negatively, then all the differences will go in one direction. If however, the treatment does not affect the subjects, some differences would go in one direction, some in the other, and some would be equal to 0. The test proceeds as follows.

1. Take the difference between the two measures.
2. Rank the differences without regard to the sign, although the sign will be needed later.
3. Add up the ranks of the negative ($\Sigma R-$) and the positive ($\Sigma R+$) differences separately. If some of the differences are 0 they should be equally divided between the positive and the negative differences. In case there is an odd number of 0 differences, one should be discarded, *thus reducing the sample size by one*, and the others should be equally divided.
4. Choose the smaller sum of the ranks T, and compare it with the critical values provided by table 7 in Appendix C. **If the sample T is less than or equal to the table's value, then the null hypothesis is rejected.** The table functions in this manner because, if the treatment affected all subjects in one direction, then all the differences would be either positive or negative. Therefore, the smaller sum of ranks will be 0, since there will be no differences which have the other sign. Hence, the smaller the value of T, the more different the two measures are.

We will practice this test with the following example: an anthropologist is interested in the permanency of prestige in a community. He interviews a member of the community involved in religious activities, who is outside the community's prestige hierarchy, and gets the informant's view of the hierarchy. The researcher asks the subject to rank-order from the highest to the lowest prestige eight heads of household. The anthropologist returns a year later, and asks the same informant to rank the same individuals. The null hypothesis is that there has been no change. This is then a two-tailed test. The data are below:

Individual ranked	Rank first year	Rank second year	Difference (first − second)
1	2	3	−1
2	7	7	0
3	8	8	0
4	1	2	−1
5	4	9	−5
6	5	4	1
7	6	5	1
8	3	4	−1

1,2. In order to compute the ranks, the differences are ordered as follows:

Ordered difference	Initial rank	Final rank
0	1	1.5
0	2	1.5
1	3	5
1	4	5
−1	5	5
−1	6	5
−1	7	5
−5	8	8

3. We add up the ranks of positive and negative differences separately:

$$\Sigma R+ = 1.5+5+5 = 11.5, \quad \Sigma R- = 1.5+5+5+5+8 = 24.5$$

4. We choose the smaller sum (in this case 11.5), and compare it with the critical value for $n=8$ at $\alpha = 0.05$ (cv = 3). We do not reject the null hypothesis, since our T is greater than the critical value.

. . . .

Practice problem 8.3

An anthropologist is interested in determining if, in a particular community, the menarcheal age has changed from one generation to the other. She decides to compare mothers and daughters, which makes this a matched design. In this community, however, the Western-style calendar is not used, making it difficult to ascertain the exact menarcheal age. The anthropologist estimates as best as possible the menarcheal age, in consultation with the subjects and other related females. Because the data are approximations, a non-parametric test should be used. Test the hypothesis that the menarcheal age has not changed.

Mothers' menarcheal age	Daughters' menarcheal age	Difference
15	13	2
14	14	0
12	11	1
14	13	1
11	11	0
16	14	2
15	13	2
16	15	1
14	12	2
11	**13**	**−2**
12	12	0
13	12	1

1,2. In order to compute the ranks, we order the differences as follows:

Ordered difference	Initial rank	Final rank
0	1	2
0	2	2
0	3	2
1	4	5.5
1	5	5.5
1	6	5.5
1	7	5.5
2	8	10
2	9	10
2	10	10
2	11	10
−2	12	10

3. We add up the ranks of positive and of negative differences separately. Since there are three 0 differences, we drop one (thus $n = 11$ now), and divide the other two:

$$\Sigma R+ = 2 + 5.5 + 5.5 + 5.5 + 5.5 + 10 + 10 + 10 + 10 = 44, \quad \Sigma R- = 2 + 10 = 12$$

4. We choose the smaller sum, namely 12, and compare it with the critical value for $n = 11$. The critical value at $\alpha = 0.05$ is 10, so we do not reject the null hypothesis. We conclude that mothers and daughters did not experience a significantly different menarcheal age.

8.7 A research example of the use of the Wilcoxon signed-rank test

In her paper titled 'Health care decisions of households in economic crisis: an example from the Peruvian highlands', Oths (1994) investigates the impact of a severe economic crisis on households' medical treatment choices. The investigation took place in a hamlet in the Northern Peruvian Andes during Peru's worst economic crisis. The author was in the unique position of measuring household expenditure at two points in time (Time 1 = August, 1998; Time 2 = November, 1998). Specifically, the author measured changes in the expenditure in three forms of medical care: traditional, lay biomedical and biomedical. The unit of analysis is the household, whose expenditures are compared at two times, making this a paired comparison. Of note, Oths mentions that the variables under analysis are not normally distributed, necessitating a non-parametric version of the paired t test. Oths reports that there is a significant 'decrease in expenditures at Time 2 for total expenditures ($p = .0002$), for home treatment ($p = .001$) and for traditional healers ($p = .006$), but not for lay biomedical ($p = .694$) and biomedical treatment ($p = .116$)' (Oths, 1994: 250). According to Oths, the use of traditional care was selectively diminished, because households were less attentive to milder illness.

Non-parametric tests with SAS/ASSIST

I am not enthusiastic about using SAS for the computation of non-parametric tests. After all, these tests are supposed to be easy to compute by hand because they are usually applied to small samples. My main problem is that, for the Mann–Whitney test, SAS produces test statistics which are transformed into z scores (as seen in the research example presented above), so they are not exactly the same as what the user computes by hand. Moreover, I was not able to find a menu-driven manner of computing a non-parametric paired comparison.

To compute either a Kruskal–Wallis or a Mann–Whitney U test, the user needs to enter the data into two columns, one of which has the data, the other of which has the classification variable (e.g. stratigraphic level or community). The menu-driven path is: data analysis, ANOVA, non-parametric ANOVA. At the non-parametric window, the user needs to specify that he/she wants Wilcoxon scores. Below is the output of the comparison of the archaeological oyster data.

```
        P A R 1 W A Y   P R O C E D U R E

     Wilcoxon Scores (Rank Sums) for Variable WEIGHT
                Classified by Variable LEVEL
             Sum of      Expected   Std Dev      Mean
   LEVEL  N   Scores     Under H0   Under H0     Score
   2      7   121.000000  105.0     19.6190003  17.2857143
   1     11   181.500000  165.0     22.2458553  16.5000000
   3     11   132.500000  165.0     22.2458553  12.0454545
             Average Scores Were Used for Ties

   Kruskal-Wallis Test (Chi-Square Approximation)
   CHISQ = 2.1708   DF = 2   Prob > CHISQ = 0.3378 ← tests H0
```

8.8 Key concepts

Non-parametric tests versus parametric tests
When should a non-parametric test be considered?
Mann–Whitney U test
Kruskal–Wallis test
Wilcoxon signed-rank test

8.9 Exercises

1. A social worker is interested in determining if the age of individuals in a nursing home is different between two ethnic groups. Because of the small size, use a Mann–Whitney U test to determine if the groups differ significantly in their ages.

Ethnic group 1	Ethnic group 2
85	88
87	90
88	86

Ethnic group 1	Ethnic group 2
83	85
87	89
86	82
79	89
83	88
82	87
86	

2. A medical anthropologist is interested in comparing the scores of three ethnic groups on a depression screening scale. She suspects that one of the ethnic groups has less depressive symptoms and is scoring differently than the others. Use a Kruskal–Wallis test to determine if the groups differ in their depression screening scale.

Group 1	Group 2	Group 3
20	10	16
17	16	15
13	14	12
15	15	18
12	13	19
9	11	14
20	14	16
14	13	17
18	12	14
16	9	10
18	11	15
11	10	14
19	13	16
9		

3. Use the data from the exercise 2 in chapter 7, performing a Kruskal–Wallis test.

4. A community center has hired a social worker with the purpose of educating the community on better eating habits. One of the main concerns of the center is the high consumption of eggs in the traditional diet of the community. The social worker chooses to work with 10 homemakers for a period of four weeks, and collects the self-reported number of eggs per weeks consumed in the household before and after the educational campaign. Test the null hypothesis that there has been no change in the self-reported number of eggs with a Wilcoxon signed-rank test.

Household	Eggs consumed per week before campaign	Eggs consumed per week after campaign
1	20	19
2	25	27

Household	Eggs consumed per week before campaign	Eggs consumed per week after campaign
3	12	12
4	15	16
5	22	21
6	20	20
7	18	17
8	19	19
9	30	30
10	16	15

9 Simple linear regression

In this chapter we introduce the statistical technique of regression analysis. This form of statistical study is more complex than the treatment given here would suggest: excluded are multiple and non-linear regression. Indeed, many second-year statistical courses will cover regression analysis only. As presented here, however, regression techniques will be found to be applicable to many research situations in the social sciences.

Regression analysis is applied to numerical data, usually continuous, although discontinuous data are also amenable to regression. The design of the analysis presents a departure from what we have covered in previous chapters: instead of comparing two or more samples, regression focuses on the relation between two variables. Moreover, there is a stated interest in explaining the behavior of the dependent variable according to the independent variable (in multiple regression, there are several independent variables). As you recall, the dependent or response variable (usually referred to as Y) is the one whose behavior we wish to understand and predict. The independent or predictor variable (usually referred to as X) is the one we use to understand and explain the behavior of the Y. The researcher manipulates or controls the independent variable, in order to observe the response of the dependent one. Thus, the main distinction between the X and Y is that the former can be controlled, or at least measured without error, by the researcher. The latter is free to vary, and is simply recorded (not manipulated). According to Draper and Smith (1981), the distinction between predictor and response variables is dependent upon the research project's purpose. A variable which at some point is used as a response may at a later point be the predictor. The purposes of regression analysis can be summarized as follows: mathematical description of the relation of Y and X, where X produces a result in Y; prediction of Y, based on knowledge of the relation between Y and X. Although we may predict the dependent variable given a value of the independent one, we do not mean that the X explicitly and physically 'causes' the Y. In other words, a regression analysis will not explain the physical, biological, or cultural mechanisms linking the two variables ('the cause'). It will simply say that X can predict the behavior of Y. An example will clarify this: let us presume that an anthropologist is studying the material items in households in order to predict the household's income in a rural community. For the sake of argument, let us also assume that this is a multiple regression exercise, in which there is one dependent variable (income), and a number of independent variables (e.g. number of electrical appliances, number of dogs in the house, size of the house, etc.). By means of regression analysis, the anthropologist finds that the independent variable which best predicts household income is the number of dogs. This conclusion does not imply that the more dogs a household has the larger its income will become. This conclusion does not say that 'X causes Y'. It simply says that the best predictor of the behavior of the Y out of a number of independent variables is that particular X.

This chapter first provides an extensive overview of regression analysis, then discusses each step of the process. Because regression analysis is a multi-step procedure instead of a one-hypothesis test, it is important to describe the entire process first.

9.1 An overview of regression analysis

In regression analysis, sample data are used to test a hypothesis about the **parametric relation** between two variables. The parameter estimate measures the strength and direction of the effect of X on Y. Regression proceeds in the following fashion.

1. Plot and inspect the data. By convention, the independent variable is plotted in the horizontal axis, and the dependent variable in the vertical axis. These axes are frequently referred to as the X and Y axes. Thus, each observation will be marked with a dot at the intersection of its X and Y values. The entire sample will form a scatter of points. If this scatter has a shape which is not linear, that is, if the scatter forms a curve, then linear relation can not be applied to the data set. Figure 9.1 shows the scatter plot of two randomly generated variables. The scatter, which shows no clear linear increasing or decreasing pattern, is what would be expected if two unrelated, random variables were plotted together. Most data points are symbolized by a '1'. If the points are symbolized by a '2', they represent two data points which had the same value.

2. Describe the relation between the X and the Y mathematically with an equation. The reader probably recalls that a straight line is described by an equation which does not have terms raised to any power (such as X^2 or X^3). Curvilinear regression (not covered here), incorporates such terms because its regression equation describes a curved line. Now, if we are to compute a linear equation that best describes a scatter of points, which line should we describe? We can trace lines across any scatter, in any direction. *The line that we want is the line that is closest to all points at once*, and is the one obtained with the 'least-squares' technique. The equation that describes this line *in the sample* will be of the form:

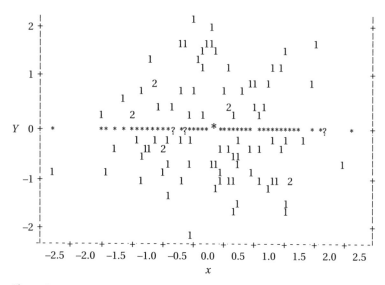

Figure 9.1. A scatter plot of unrelated variables and their regression equation.

....

Formula 9.1 Sample formula of the regression equation

$$\hat{Y} = a + b(X) + \epsilon$$

where $\hat{Y}$ (Y hat) is the predicted value of Y, given the mathematical relation
between Y and X described by the equation,
a is the point at which the line intercepts the Y axis, and
b is the slope of the line. When we write down $b(X)$, we mean that the slope,
multiplied by a particular X value, can predict the value of Y, in conjunction with the rest of the equation.
ϵ is the error (lack of precision) involved in the prediction and explanation of
Y.

....

We know something else about this line: it must go through the means of
both variables $\overline{X}, \overline{Y}$. This is a useful tip, when drawing the line by hand. The
means of the variables plotted in figure 9.1 are:

Variable	Mean
X	0.0811784
Y	0.0036238

The line crosses the point at which the value of both means intercepts,
which has been marked by a bold asterisk. The regression line obtained for
the two randomly generated, unrelated variables is shown on figure 9.1 as a
line of small asterisks. There is a '?' at points in which the line was super-
imposed with the actual data points.

3. *Express the regression analysis as an analysis of the variance of Y.* Using
the same computations already used to compute the equation, we are able to

compute sums of squares (SS) and mean squares (MS), which quantify the proportion of the total variation of Y which can be explained by X (SSregresssion) and the proportion which remains un-explained (SSerror). The mean squares as always are obtained by dividing the SS by their df. Using the MS, an F ratio is computed to test the null hypothesis that X does not explain a significant amount of Y's variation. **The coefficient of determination or R^2**, which expresses in a scale of 0 to 1 the proportion of the total variation of Y explained by X, is also computed. If $R^2 = 0$, then X does not explain any of the variation of the Y, and if it is equal to 1, X explains all the variation of the Y. Notice that the R^2 does not test a hypothesis.

4. *Test the null hypothesis that the **parametric** value of the slope is not statistically different from 0.* The component of the sample equation which expresses the strength and direction of the relation between the X and Y is the slope. The Y intercept only shows at which point the line will cross the Y axis, but does not inform us about the statistical significance of the relation between the variables. However, if b is positive, as X increases Y increases. If it is negative, as X increases Y decreases. And if b is not statistically significantly different from 0, then the X does not affect, and can not be used as a predictor for, the behavior of Y.

Although we will not be working with populations, the parametric formula is shown below. The parametric regression equation uses, as is always done with parameters, Greek symbols:

....

Formula 9.2 Population formula of the regression equation

$$Y = \alpha + \beta(X)$$

where Y is known without error, hence the lack of the $\wedge$ on top of it, and the lack of the error term,

α is the parametric value at which the line described by the equation intersects the Y axis,

β is the slope of the line. When we write down $\beta(X)$, we mean that the slope, multiplied by a particular X value, results in the known value of Y, in conjunction with the rest of the equation.

....

5. *Use the regression equation to predict values of Y.* This is done by simply substituting in the regression equation the sample values for a, b and for the X whose Y we want to predict. The values are usually given with confidence intervals, since the confidence associated with our predictions depends on the value of X entered in the equation. If the value of X is $\overline{X}$, or very close to it, our confidence is greatest and the confidence intervals are smallest. If the value of X is very different from $\overline{X}$, our confidence in the prediction is diminished, and the intervals are wider. It is customary at this point to present a plot showing: the observed data, the equation line, and the confidence intervals.

6. *Analyze the residuals.* The question at this point is: how well does the model 'fit' the data? If the fit is poor, what can be done to improve it? The two usual responses to the last question are: transform the data to make it linear, or use non-linear regression.

We will now cover all these steps with part of a data set I collected in Limon, Costa Rica. The data consist of various reproductive variables such as age at menarche, age at first birth, total number of pregnancies, number of miscarriages/abortions/stillbirths, number of livebirths, age at onset of menopause, and total number of surviving offspring at the time of the interview. All females were at least 50 years of age, the age at which reproduction is said to stop in human females. All subjects said they never used Western-style contraception (or any other form, for that matter). If we are interested in explaining and predicting achieved fertility in this group, we could use a multiple regression analysis, in which we incorporate all or several of these variables. However, let us take a similar problem by using only two variables: the total number of pregnancies as the X, and the achieved fertility (number of surviving offspring at time of the interview) as the Y. We ask the following questions:

1. Does the number of pregnancies affect and predict achieved fertility?
2. How can we describe mathematically the relationship between these two variables (at this point, we would also wish to compute predicted values with their confidence intervals)?
3. Does the number of pregnancies explain a significant amount of the variation in achieved fertility?
4. What proportion of the variation of the Y is explained by the X?
5. Does the mathematical model described by the regression equation fit the data? Is there a better model which could be achieved by transforming the data?

For this exercise, we will use only 37 females, whose data are as follows:

Total number of pregnancies	Surviving offspring at time of interview
8	9
8	6
8	7
5	5
9	8
1	0
2	2
4	4
6	7
11	10
5	5
5	5

Total number of pregnancies	Surviving offspring at time of interview
3	3
6	5
4	4
1	1
1	0
0	0
6	6
5	5
7	5
14	14
3	3
13	12
8	8
3	3
9	8
11	11
8	6
9	8
11	8
3	3
3	3
6	5
1	1
16	15
3	3

9.2 Plot and inspection of the data

Using SAS/ASSIST, a plot was obtained which showed the X on the horizontal line, the Y on the vertical axis. If two females had the same data, then they are represented by a '2', and so on. The plot is shown in figure 9.2.

The plot shows that there is a clear linear, *non-curved* relation between the X and Y. However, there seems to be more variation in the middle range of the data set.

9.3 Description of the relation between X and Y with an equation

We are now going to compute the regression equation, using the sample formula, presented in formula 9.1: $\hat{Y} = a + b(X) + \epsilon$. To compute the equation, we need eight quantities: $\bar{Y}, \bar{X}, n, \Sigma X, \Sigma X^2, \Sigma Y, \Sigma Y^2$, and ΣXY. For our data set:

$\bar{Y} = 5.62, \bar{X} = 6.11, n = 37, \Sigma X = 226, \Sigma X^2 = 1{,}934, \Sigma Y = 208, \Sigma Y^2 = 1{,}660$

and $\Sigma XY = 1{,}780$

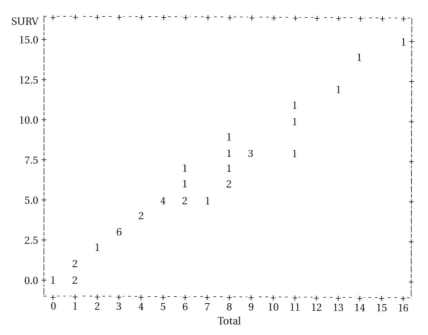

Figure 9.2. A scatter plot of the X and the Y variables, and their regression equation.

We follow these steps.

1. Compute the sums of squares of X, which will be referred to as SS_X. For the reader's reference, the formula for the sum of squares is shown here, but not in the rest of the chapter:

$$SS_X = \Sigma X^2 - \frac{(\Sigma X)^2}{n} = 1{,}934 - \frac{(226)^2}{37} = 1{,}934 - \frac{51{,}076}{37} = 1{,}934 - 1{,}380.43 = 553.57$$

2. Compute the sum of squares of Y, which will be referred to as SS_Y:

$$SS_Y = 1{,}660 - \frac{(208)^2}{37} = 1{,}660 - \frac{43{,}264}{37} = 1{,}660 - 1{,}169.3 = 490.70$$

3. A new quantity commonly referred to as the sum of products is now introduced. This quantity is similar to the sum of squares, but considers the variation of both Y and X. The sum of products will be referred to as SP_{XY}, and is computed as follows:

Formula 9.3 The sum of products (SP_{XY})

$$SP_{XY} = \Sigma XY - \frac{(\Sigma X)(\Sigma Y)}{n}$$

In our data set, SP is

$$1,780 - \frac{(226)(208)}{37} = 1,780 - \frac{47,008}{37} = 1,780 - 1,270.78 = 509.51$$

4. Compute the slope, according to the following formula:

Formula 9.4 The regression slope

$$b = \frac{\text{SP}}{\text{SS}_X}$$

In our example,

$$b = \frac{509.51}{553.57} = 0.92$$

5. The last step in the computation of the regression equation is the Y intercept, that is, the place in the Y axis at which the line intercepts. The formula for the computation of the Y intercept is:

Formula 9.5 The regression line Y intercept

$$a = \overline{Y} - (b)(\overline{X})$$

In our example, $\overline{Y} = 5.62$, $\overline{X} = 6.11$. Thus, $a = 5.62 - (0.92)(6.11) = 5.62 - 5.6212$ $= -0.0001$. Therefore, the sample regression equation in our sample is:

$$\hat{Y} = -0.0001 + 0.92(X) + \epsilon$$

By looking at this equation, we know that the line will cut the Y axis just under 0. We can also see that the relation between X and Y is positive (that is, as the number of pregnancies increases, so does the number of surviving children). We next test if such a relation is significant.

9.4 Expression of the regression analysis as an analysis of variance of Y

A complete regression analysis should include an ANOVA of the dependent variable. The purpose here is to partition the variation of the Y, and determine:

- whether X explains a significant amount of variation of the Y. This is indeed a null hypothesis (which is not written in parametric terms), and is tested with an F ratio.
- what proportion of the total variation of the Y can be explained by the X.

We are therefore going to construct an analysis-of-variance table which will have two sources of variation: regression (the regression model, or the independent variables) and error (that part of the variation of the *Y* which can not be explained by the *X*). The table will show three sums of squares: SStotal, SSreg, SSerror. By definition, these sums of squares are additive. The table will also have degrees of freedom associated with the SSreg and the SSerror. As such, they will be referred to as regression and error df. We will also compute the mean square due to regression (MSreg) by dividing the SSreg by its df, and the mean square due to error (MSerror) by dividing the SSerror by its df. We will compute an *F* ratio by dividing the MSreg by the MSerror, and compare it with the table of *F* ratios, using the associated df to find the critical value. If the test *F* ratio is greater than or equal to the critical value at $\alpha = 0.05$, the null hypothesis that *X* does not explain a significant part of the variation of *Y* is rejected. Fortunately, the computations necessary for constructing this table have already been obtained when the regression equation was computed. The ANOVA table, with its formulae, is shown below:

Formulae 9.6 Formulae for computation of ANOVA in a *simple* regression analysis

Source	df	SS	MS	F
Regression	1	$\dfrac{(\mathrm{SP}_{XY})^2}{\mathrm{SS}_X}$	$\dfrac{\mathrm{SSreg}}{1}$	$\dfrac{\mathrm{MSreg}}{\mathrm{MSerror}}$
Error	$n-2$	$\mathrm{SStotal}-\mathrm{SSreg}$	$\dfrac{\mathrm{SSerror}}{n-2}$	
Total	$n-1$	SS_Y		

In our example, we already computed the following:

degrees of freedom: $n = 37$, so $n-1 = 36$ and $n-2 = 35$
sums of squares: $\mathrm{SS}_Y = 490.70$, $\mathrm{SS}_X = 553.57$, $\mathrm{SP}_{XY} = 509.51$.

Therefore, $\mathrm{SStotal} = \mathrm{SS}_Y = 490.7$. The SSreg is thus

$$\mathrm{SSreg} = \frac{(509.51)^2}{553.57} = \frac{529{,}600.44}{553.57} = 468.95.$$

Since the sums of squares are additive, the SSerror is computed by subtraction: $\mathrm{SSerror} = 490.70 - 468.95 = 21.75$. The MSreg is $= 468.95/1 = 468.95$. In a simple regression ANOVA table, since the regression df are always 1, the MSreg = SSreg. In a multiple regression equation, the regression df will vary according to the number of independent variables, so the MSreg and the SSreg will not be equal. The MSerror is $= 21.75/35 = 0.62$. Finally, we compute our *F* ratio by dividing the MSreg by the MSerror: $F = 468.95/0.62 = 756$. The table does not list the critical values at our df, namely 1 and 35. We therefore use the more conservative values at 1 and 30. Clearly, our *F* ratio is highly significant, since it far exceeds the critical value at 0.001 (cv = 13.3).

We therefore reject the null hypothesis that the total number of pregnancies does not explain a significant amount of the variation of Y. The next step is to quantify exactly how much variation is being explained. We first present the computations in an ANOVA table:

ANOVA table of the fertility regression analysis

Source	df	SS	MS	F	p
Regression	1	468.95	468.95	756	<0.001
Error	35	21.75	0.62		
Total	36	490.70			

To quantify what proportion of the total variation of Y is explained by the X, we compute **the coefficient of determination or** R^2 which expresses in a scale of 0 to 1 the proportion of the total variation of Y explained by X. SAS (SAS Institute Inc. SAS/STAT User's Guide, Release 6.03 Edition. Cary, NC: SAS Institute Inc. 1988) also computes by default a statistic called the adjusted R^2 (R^2 adj), in which the number of independent variables in the model is considered. In a simple regression analysis, the R^2 and the R^2 adj are not likely to be very different. However, in a multiple regression analysis, if a not-too-wise user were to feed the computer a large number of independent variables, and made it compute a regression equation, he/she could obtain a large, *inflated* R^2, while at the same time obtaining a low R^2 adj. The latter statistic is simply more conservative, and should be preferred over the usual R^2. The formulae for both are shown below:

....

Formula 9.7 Formulae for the R^2 and the R^2 adj for a *simple* regression model

$$R^2 = 1 - \frac{\text{SSreg}}{\text{SStotal}}$$

$$R^2 \text{ adj} = 1 - \frac{(n-1)(1-R_9)}{n-2}$$

....

We first compute the coefficient of determination:

$$R^2 = 1 - \frac{468.95}{490.7} = 1 - 0.9557 = 0.9557$$

Then,

$$R^2 \text{ adj} = 1 - \frac{(36)(1-0.9557)}{35} = 1 - \frac{1.59}{35} = 1 - 0.04556 = 0.9544$$

9.5 Test of the null hypothesis H0: $\beta=0$

If figure 9.1 is inspected, it will be noticed that the equation line between two randomly generated and totally unrelated variables is flat. That is, the slope of such a line is 0, or very close to 0. Hence our null hypothesis: if the slope is not significantly different from 0, then the X cannot explain or predict the behavior of the Y. The null hypothesis test proceeds as follows: the sample slope (b) is used to estimate β, and to test H0: $\beta=0$. The test statistic will be a t score with df $= n-2$. This t score does not differ from those used in previous chapters: In the numerator, it has a difference, in this case between the slope computed from the sample data, and the slope proposed by the null hypothesis. Since the latter is 0, it is dropped from the numerator. In the denominator, the t score formula incorporates a measure of error or variation. In this case it is a standard error that includes the portion of the variation of the Y which remains unexplained (MSerror) and a measure of the variation of the X (SS$_X$). If the H0 is not rejected, the study is not continued, since there is no evidence of a significant association between the variables. If the null hypothesis is rejected, the regression analysis may continue. The formula for the computation of the t statistic uses statistics which have already been computed, and is shown below:

Formula 9.8 The t score for H0: $\beta=0$

$$t=\frac{b-0}{S_b}$$

$$\text{where } S_b=\sqrt{\frac{\text{MSerror}}{\text{SS}_X}}$$

In our example, we know that the MSerror $=0.62$, and that SS$_X=553.57$. Therefore:

$$S_b=\sqrt{\frac{0.62}{553.57}}=\sqrt{0.00112}=0.033, \quad t=\frac{0.92-0}{0.033}=27.5, \quad df=35$$

The critical value at df $=35$ is not listed, so we use the more conservative values at df $=30$. We reject the null hypothesis with great confidence ($p<0.001$).

9.6 Use of the regression equation to predict values of Y

When predicted values of Y are computed, the meaning of the value of the slope can be grasped: the slope is the amount by which Y will be increased, if the value of the X is increased by 1. Although the complete equation is

$\hat{Y} = -0.0001 + 0.92(X) + \epsilon$, the error term is not considered for computation purposes. The researcher simply acknowledges that there is error in the calculation, and accompanies the prediction with confidence intervals. Let us first compute predicted values of Y for the following values of X: 1, 2, 5, 6, 10 and 11.

X	$\hat{Y} = -0.0001 + 0.92(X)$	$\hat{Y}$
1	$-0.0001 + 0.92(1)$	0.92
2	$-0.0001 + 0.92(2)$	1.84
5	$-0.0001 + 0.92(5)$	4.6
6	$-0.0001 + 0.92(6)$	5.52
10	$-0.0001 + 0.92(10)$	9.2
11	$-0.0001 + 0.92(11)$	10.12

Notice that $1.84 - 0.92 = 0.92$, $4.6 - 1.84 = 0.92$, and $10.12 - 9.2 = 0.92$, which is the value of the slope. Thus, if we increase the value of X by 1, say, from 10 to 11, the predicted value of the Y will be increased by 0.92.

When reporting prediction results, it is usual to accompany the prediction with some sort of interval, since the prediction cannot be totally precise. Two confidence intervals (CI) are covered here. Two forms of predictions can be made with regression: one is for an individual $\hat{Y}$ for a specific X. The second interval requires more explanation: for each value of X there is a population of Y values. For example, for a female with X number of pregnancies, there is a population of possible values of Y. Indeed, we see variation in the observed Y value of females with the same X value. It is possible to predict the parametric mean of a population of $\hat{Y}$ associated with a specific X by using regression as done previously. However, the confidence intervals are different. Thus, whereas the first CI surrounds an individual $\hat{Y}$, the second one surrounds the parametric mean $\mu_{\hat{Y}}$ associated with that $\hat{Y}$.

First we cover the confidence intervals for one individual $\hat{Y}$:

Formula 9.9 Confidence interval (CI) for an estimated $\hat{Y}$ for a given value of X

$$95\% \text{ CI for an individual } \hat{Y} = \hat{Y} \pm (S_{\hat{y}})(t_{0.05,\text{df}})$$

$$\text{where } S_{\hat{y}} = \sqrt{\text{MSerror}\left[1 + \frac{1}{n} + \frac{(X_i - \overline{X})^2}{SS_X}\right]}$$

$S_{\hat{y}}$ is the standard error for the individual $\hat{Y}$, X_i is the X for which we are predicting the $\hat{Y}$, and $(t_{0.05,\text{df}})$, is the critical value in the t table at $\alpha = 0.05$ for df $= n - 2$. The $\pm$ symbol means that we subtract from and add to the $\hat{Y}$ the product of the standard error by the t score from the table.

Formula 9.10 Confidence interval for $\mu_{\hat{Y}}$, the parametric value associated with $\hat{Y}$

$$95\% \text{ CI for } \mu_{\hat{Y}} = \hat{Y} \pm (S_{\mu_{\hat{Y}}})(t_{0.05,\text{df}})$$

$$\text{where } S_{\mu_{\hat{Y}}} = \sqrt{\text{MSerror}\left[\frac{1}{n} + \frac{(X_i - \overline{X})^2}{SS_X}\right]}$$

We proceed to compute both standard errors and confidence intervals for a few of the predicted values already computed. We know that MSerror $= 0.62$, $n = 37$, $\overline{X} = 6.11$, $SS_X = 553.57$, and that the t value at $\alpha = 0.05$ with df $= 35 = 2.03$ (this value was obtained by extrapolation, since it is not in the table; for information about extrapolation, see Sokal and Rohlf, 1981). We first begin with *the confidence interval for an individual $\hat{Y}$*.

$$X \quad \hat{Y} \qquad S_{\hat{Y}} = \sqrt{\text{MSerror}\left[1 + \frac{1}{n} + \frac{(X_i - \overline{X})^2}{SS_X}\right]} \quad \begin{array}{l}95\% \text{ CI for } \hat{Y} \\ = (S_{\hat{Y}})(t_{0.05,\text{df}})\end{array}$$

$$1 \quad 0.92 \qquad 0.8162 = \sqrt{0.62\left[1 + \frac{1}{37} + \frac{(1 - 6.11)^2}{553.57}\right]} \quad \begin{array}{l}\text{CI for } 0.92 \\ = (0.8162)(2.03) = 1.657 \\ 0.92 + 1.657 = 2.57 = \text{upper} \\ 0.92 - 1.657 = -0.737 = \text{lower}\end{array}$$

$$6 \quad 5.52 \qquad 0.798 = \sqrt{0.62\left[1 + \frac{1}{37} + \frac{(6 - 6.11)^2}{553.57}\right]} \quad \begin{array}{l}\text{CI for } 5.52 \\ = (0.798)(2.03) = 1.62 \\ 5.52 + 1.62 = 7.14 = \text{upper} \\ 5.52 - 1.62 = 3.90 = \text{lower}\end{array}$$

$$11 \quad 10.12 \quad 0.815 = \sqrt{0.62\left[1 + \frac{1}{37} + \frac{(11 - 6.11)^2}{553.57}\right]} \quad \begin{array}{l}\text{CI for } 10.12 \\ = (0.815)(2.03) = 1.65 \\ 10.12 + 1.65 = 11.77 = \text{upper} \\ 10.12 - 1.65 = 8.47 = \text{lower}\end{array}$$

Notice that the standard error diminishes when X is close to the mean, whereas it increases if X is farther away from it. Next, the confidence interval for the parametric value of the mean associated with the particular $\hat{Y}$ is computed.

$$X \quad \hat{Y} \qquad S_{\mu_{\hat{Y}}} = \sqrt{\text{MSerror}\left[\frac{1}{n} + \frac{(X_i - \overline{X})^2}{SS_X}\right]} \quad \begin{array}{l}95\% \text{ CI for } \mu_{\hat{Y}} \\ = (S_{\mu_{\hat{Y}}})(t_{0.05,\text{df}})\end{array}$$

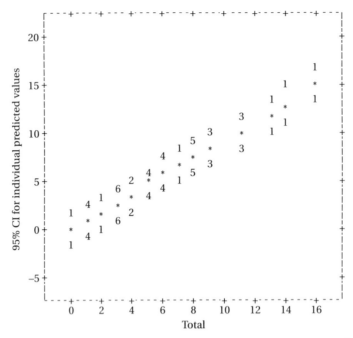

Figure 9.3. A plot of the equation line and the confidence intervals for $\hat{Y}$.

$$1 \quad 0.92 \qquad 0.214 = \sqrt{0.62\left[\frac{1}{37} + \frac{(1-6.11)^2}{553.57}\right]} \quad \text{CI for } 0.92$$
$$= (0.214)(2.03) = 1.435$$
$$0.92 + 0.435 = 1.355 = \text{upper}$$
$$0.92 - 0.435 = 0.485 = \text{lower}$$

$$6 \quad 5.52 \qquad 0.13 = \sqrt{0.62\left[\frac{1}{37} + \frac{(6-6.11)^2}{553.57}\right]} \quad \text{CI for } 5.52$$
$$= (0.13)(2.03) = 0.263$$
$$5.52 + 0.263 = 5.783 = \text{upper}$$
$$5.52 - 0.263 = 5.257 = \text{lower}$$

$$11 \quad 10.12 \qquad 0.209 = \sqrt{0.62\left[\frac{1}{37} + \frac{(11-6.11)^2}{553.57}\right]} \quad \text{CI for } 10.12$$
$$= (0.209)(2.03) = 0.424$$
$$10.12 + 0.424 = 10.54 = \text{upper}$$
$$10.12 - 0.424 = 9.7 = \text{lower}$$

Once again, the width of the interval decreases if X is close to the mean. Thus, if we wish to predict either an individual value, or the value of the parametric mean associated with said individual value, our confidence is diminished the farther away we move from the mean of the independent variable.

This section ends with a plot of the equation line and the confidence intervals for $\hat{Y}$ (figure 9.3). There is a topic that on the surface is rather obvious, and is thus frequently left out in statistics textbooks. That is, the line which is plotted consists of the predicted values generated by the equation. Therefore, note that, for the X values of 1, 6, and 11, the line has the $\hat{Y}$ values we computed, namely: 0.92, 5.52 and 10.12. Notice that the line crosses the Y axis at the point we calculated, that is, at $a = -0.0001$. Also note that the line crosses the point of the intersection of the two variables' means: $\overline{Y} = 5.62$, $\overline{X} = 6.11$. A plot with both confidence intervals would have been too cumbersome, and one is enough to show that the interval's width increases as the value of the X becomes more and more different from $\overline{X} = 6.11$. If there is more than one item at the same place, SAS places a '?' mark. Otherwise, the equation line is represented by a '*'.

9.7 Residual analysis

After the regression equation has been computed, we have for each observation its observed X and Y values, and its $\hat{Y}$. If the predicted value is subtracted from Y, then we obtain ϵ, the error in our prediction. This is the portion of Y's variation that remains un-accounted for in regression analysis. It is of extreme importance that the residuals be plotted against the X because they will reveal if the regression model is not appropriate. If the model fits, the scatter of the residuals should present no distinct shape. If, however, the residuals have a curved shape, or if the residuals look like a fan, then the data may have to be transformed or a non-linear model may have to be applied. Please note that, even if a model is found to be significant, it could still suffer from lack of fit. For example, if the relation between X and Y is curvilinear, and a linear regression is applied to the data, the model could very well be significant because it explains the portion of X and Y's relation which is linear. However, upon examination of the residuals, the researcher would notice that they form a curve, indicating that the curved aspect of the variables' relation has not been included in the regression equation. If the data are transformed and made to have a linear relation, the R^2 of the model would likely increase. A 'fanning effect' suggests that the data are more variable on one end than on another. In figure 9.4, the residuals from the regression analysis of the fertility data are shown. Note that there is no clear curved pattern to the scatter although the residuals tend to have higher values for the mid range of the X. This is not surprising, since we already observed greater variation in the middle range of the data.

If the residuals indicate that a transformation is necessary, then the user should try first to transform the X variable. Commonly used transformations are the log, the arcsine, the square root of X, and X^2, X^3, etc. It is best to try all these transformations before attempting to transform the Y, which as the dependent variable is not under the control of the investigator. For an excellent review of transformations in regression analysis, see Draper and Smith (1981).

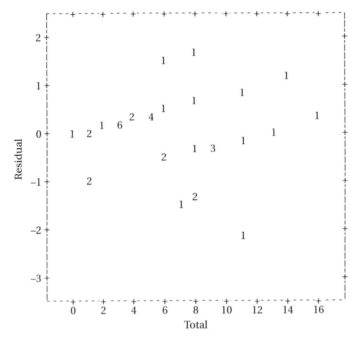

Figure 9.4. A plot of the residuals against the independent variable.

Another easily performed residual analysis is to test the null hypothesis that the residuals are normally distributed. That is, after the variation of the dependent variable attributable to the independent variable has been accounted for by the regression analysis, the remaining variation of the Y should have a normal distribution, just like a random data set would. It is extremely easy to save the residuals for further analysis with SAS. In the fertility example, the residuals were saved, and a PROC UNIVARIATE NORMAL analysis was performed. For the sake of space, only part of the output is reproduced:

```
                    Analysis Variable : R Residual
                    Univariate Procedure

                        Moments
N                       37   Sum Wgts          37
Mean                     0   Sum                0
Std Dev           0.777055   Variance    0.603815
Skewness         -0.57244    Kurtosis    1.022907
USS               21.73733   CSS         21.73733
CV                       .   Std Mean    0.127747
T:Mean=0                 0   Pr>|T|        1.0000
Num ^=0                 37   Num>0             24
M(Sign)                5.5   Pr>=|M|       0.0989
Sgn Rank              36.5   Pr>=|S|       0.5886
W:Normal          0.940215   Pr<W          0.0620  ←very close to 0.05
```

We can see that, although the p value of the W statistic is 'legally' non-significant, it approaches the critical value. We go a step further, and obtain a histogram of the residuals, which is shown in figure 9.5.

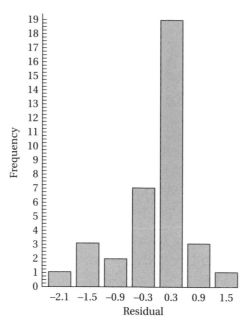

Figure 9.5. A histogram of the residuals.

The histogram confirms that the distribution of the residuals is not normal, although the p value of the normality test is greater than 0.05. SAS also offers an excellent analysis of residuals (see figure 9.6): the residuals are divided by their standard error to produce the so-called studentized residual, printed and plotted. Residuals with very large numbers have a large influence, and could point to an error in data entry. Cook's D, a measure of the residual's influence, is also printed. The entire output is not shown, only the plot and Cook's D:

Two residuals stand out: those of observations #22 and #31. These observations are influential because their predicted and observed values are quite different. Observation # 22 had 14 pregnancies, and at the time of the interview she had 14 living offspring, whereas she was predicted to have 12.88. Observation #31 had 11 pregnancies, and at the time of the interview she had eight offspring alive, whereas she was expected to have 10.12. Therefore, we can conclude that these observations had large residuals not because of errors in data entry, but because of actual variation in the data set. Although the model we applied to the data explains a great amount of the variation of the Y, and its residuals are acceptable, perhaps it would be best if the X were transformed. This is the benefit of having used a 'real-life' example: data do not always easily conform to textbooks' requirements! The studious reader is encouraged to try to improve the model. Transformations are exceedingly easy to perform with the help of computers, and need not be covered any more in this book.

Obs	−2−1−0 1 2	Cook's D
1	****	0.077
2	***	0.054
3		0.004
4	*	0.004
5		0.003
6	**	0.059
7		0.001
8		0.003
9	***	0.050
10		0.001
11	*	0.004
12	*	0.004
13		0.002
14	*	0.006
15		0.003
16		0.000
17	**	0.059
18		0.000
19	*	0.005
20	*	0.004
21	***	0.051
22	***	0.188 ◄——— large residual
23		0.002
24		0.000
25	*	0.012
26		0.002
27		0.003
28	**	0.050
29	***	0.054
30		0.003
31	*****	0.295 ◄——— large residual
32		0.002
33		0.002
34	*	0.006
35		0.000
36		0.019
37		0.002

Sum of residuals	0
Sum of Squared Residuals	21.7373
Predicted Resid SS (Press)	24.2218

Figure 9.6. Output of the residual analysis produced by SAS.

Practice problem 9.1

An anthropologist is interested in determining if the number of members in a household impacts the nutritional status of its young children. Although it would be best to record the actual nutrient consumption of the children, the anthropologist decides to measure the weight of a group of children from a homogeneous community who have been matched for age (9 years), gender, socio-economic status, occupation, religion, etc. Thus, the null hypothesis is that the number of members in a household does not affect the weight of the children. The data (as entered in SAS) are shown below:

```
Number of
individuals    Weight
in household   in kilos
     2            33
     6            28
     8            27
     3            35
    10            25
     7            27
     2            32
     6            29
     9            27
     5            28
     6            27
     4            31
     5            26
     7            25
     3            32
      n = 15
```

We follow the same steps we took previously. The formulae, however, are not reproduced here.

1. *Plot and inspect the data.* The plot indicates that there is a linear, negative relation between the X and the Y. No curved relation is visible.

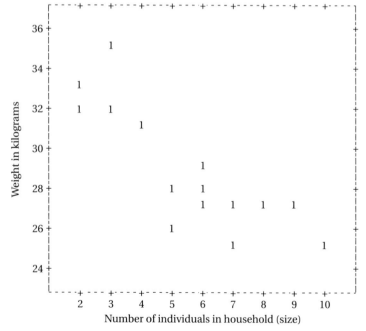

Plot of the X and Y variables.

2. *Describe the relation between the X and the Y mathematically with an equation.* To compute the equation, we need eight quantities:

$$\bar{Y}=28.8,\ \bar{X}=5.53,\ n=15,\ df=12,\ \Sigma X=83,\ \Sigma X^2=543,\ \Sigma Y=432,\ \Sigma Y^2=12{,}574,$$
$$\Sigma XY=2{,}302$$

With these quantities, we proceed to compute the SS_X, the SS_Y, and the SP_{XY}:

$$SS_X=543-\frac{(83)^2}{15}=543-459.27=83.73$$

$$SS_Y=12{,}574-\frac{(432)^2}{15}=12{,}574-12{,}441.6=132.4$$

$$SP_{XY}=2{,}302-\frac{(83)(432)}{15}=2{,}302-2{,}390.4=-88.4$$

We compute the slope and the intercept, where

$$b=\frac{-88.4}{83.73.}=-1.055,\quad \text{and } a=28.8-(-1.055)(5.53)=34.63$$

Therefore, the regression equation in our sample is the following: $\hat{Y}=34.64-1.055(X)+\epsilon$. We can tell at this point that the line will cross the Y axis above the mean of Y, and that it has a negative slope. That is, the more members in a household, the lower the children's weight. We still do not know if this relation is significant.

3. *Express the regression analysis as an analysis of the variance of Y.* First we compute the sums of squares. The SStotal is the $SS_Y=132.4$. The SSregression is:

$$SSreg=\frac{(-88.4.)^2}{83.73}=93.33.$$

Since the sums of squares are additive, we obtain the error SS by subtraction: $SSerror=132.4-93.33=39.07$. The degrees of freedom associated with the SSreg are 1, and those associated with the SSerror are $15-2=13$. Thus,

$$MSreg=\frac{93.33}{1}=93.33,\quad \text{and } MSerror=\frac{39.07}{13}=3.00.$$

The F ratio is computed as $F=93.33/3=31.1$, $df=1,\ 13$. Our F ratio is highly significant ($p<0.001$, since the critical value is 17.8 at $\alpha=0.001$). We can reject the null hypothesis that the household size does not explain the variation of children's weight. We display our results in the usual ANOVA table, including both the R^2 and the R^2 adjusted:

ANOVA table

Source	df	SS	MS	F	p
Regression	1	93.33	93.33	31.1	<0.001
Error	13	39.07	3.00		
Total	14	132.4			

$$R^2=1-\frac{39.07}{132.4}=0.705,\quad R^2\ adj=1-\frac{(14)(1-0.705)}{13}=0.682.$$

Thus, the X explains about 68% of the behavior of the Y.

4. *Test the null hypothesis H0: $\beta = 0$.* First, we compute the standard error of the slope:

$$S_b = \sqrt{\frac{3}{83.73}} = 0.189, \quad t = \frac{-1.055}{0.189} = -5.57, \text{ df} = 13$$

Our t score is much higher than the critical value at $\alpha = 0.001$. Therefore, we reject the null hypothesis that the slope is not different from 0.

5. *Use the regression equation to predict values of Y.* Let us first compute predicted values of Y, and the confidence intervals for the individual prediction $\hat{Y}$ for the following values of X: 3 and 7. The t value we will use from the table is $t_{0.05,13} = 2.16$. The MSerror, taken from the ANOVA table, is 3.

X	$\hat{Y} = 34.64 - 1.055(X)$	$S_{\hat{Y}}$	95% CI for $\hat{Y}$
3	$31.475 = 34.64 - 1.055(3)$	$1.85 = \sqrt{3\left[1 + \frac{1}{15} + \frac{(3-5.53)^2}{83.73}\right]}$	$(1.85)(2.16) = 4$ 31.475 ± 4 $= \begin{cases} 35.47 \text{ upper} \\ 27.46 \text{ lower} \end{cases}$
7	$27.25 = 34.64 - 1.055(7)$	$1.81 = \sqrt{3\left[1 + \frac{1}{15} + \frac{(7-5.53)^2}{83.73}\right]}$	$(1.81)(2.16) = 3.9$ 27.25 ± 3.9 $= \begin{cases} 31.16 \text{ upper} \\ 23.35 \text{ lower} \end{cases}$

Next we compute the standard error of the parametric mean associated with the individual $\hat{Y}$, and the associated confidence intervals.

X	$\hat{Y}$	$S_{\mu\hat{Y}}$	95% CI for $\mu_{\hat{Y}}$
3	31.475	$0.656 = \sqrt{3\left[\frac{1}{15} + \frac{(3-5.53)^2}{83.73}\right]}$	$(0.656)(2.16) = 1.42$ 31.475 ± 1.42 $= \begin{cases} 32.89 \text{ upper} \\ 30.05 \text{ lower} \end{cases}$
7	27.25	$0.527 = \sqrt{3\left[\frac{1}{15} + \frac{(7-5.53)^2}{83.73}\right]}$	$(0.527)(2.16) = 1.14$ 27.25 ± 1.14 $= \begin{cases} 28.38 \text{ upper} \\ 26.11 \text{ lower} \end{cases}$

Last, we plot the regression line with the confidence intervals for the individual predicted values. Once again, the line is represented by an '*':

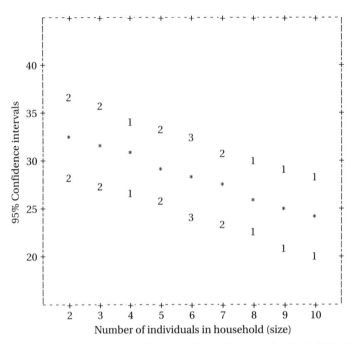

Plot of the regression line with the confidence intervals for the individual
predicted values.

6. *Analyze the residuals.* As we did previously, we first plot the residuals and look for any evidence for curved or 'fan-like' behavior:

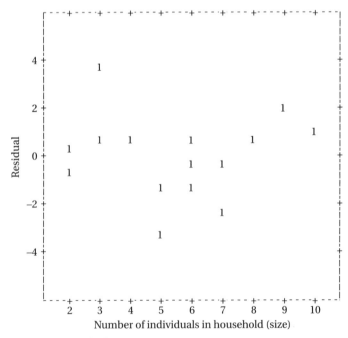

A plot of the residuals.

The data do not appear to have any consistent behavior. We, however, plot the residuals in a histogram, and test their distribution for departures from normality, as done previously:

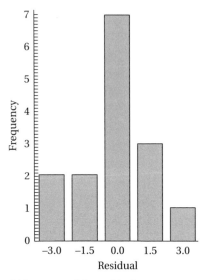

A histogram of the residuals.

```
       Univariate Procedure.  Variable = Residual
                    Moments
N                       15   Sum Wgts           15
Mean                     0   Sum                 0
Std Dev           1.670613   Variance     2.790946
Skewness          -0.03704   Kurtosis     0.859805
USS               39.07325   CSS          39.07325
CV                       .   Std Mean      0.43135
T:Mean = 0               0   Pr > |T|       1.0000
Num ^ = 0               15   Num > 0             8
M(Sign)                0.5   Pr > = |M|     1.0000
Sgn Rank                 4   Pr > = |S|     0.8469
W:Normal          0.966479   Pr < W         0.7659  ←normal data
```

Although the histogram does not show a normal distribution, the *W* statistic is clearly not significant. Last, we ask SAS to do a residual analysis, which does not indicate that any residual merits further investigation because of undue influence:

```
                                              Cook's
          Obs       -2-1-0  1 2                  D

           1       |          |               0.013
           2       |          |               0.001
           3       |        * * * *           0.020
           4       |        *                 0.404
           5       |          |               0.088
           6       |          |               0.001
           7       |          |               0.016
           8       |          |               0.006
           9       |        * *               0.194
          10       |    *     |               0.025
          11       |    *     |               0.023
          12       |          |               0.006
          13       | * * * *  |               0.152
          14       |    * *   |               0.095
          15       |          |               0.009

     Sum of residuals                        0
     Sum of Squared Residuals          39.0732
     Predicted Resid SS (Press)        50.7867
```

Output of the residual analysis produced by SAS.

Regression analysis using SAS/ASSIST

To perform regression analysis, the data need to be entered into two columns, one of which has the X, the other the Y. It does not matter which is entered first. The path to perform a regression analysis is: data analysis, regression, linear regression. At the regression window, the user is asked to specify the dependent and independent variables. At this window, the user can also click on the 'additional options' buttons. Here is where the user requests the two 95% confidence intervals we computed, plus the residual analysis, and where the user can request and even create graphs. Following this path, and requesting a residual analysis plus the two 95% confidence intervals, the example of the regression on household size of children's height is shown below. Notice that SAS also tests the null hypothesis that the Y intercept is equal to 0. This hypothesis is generally of no interest for the purpose of regression analysis. The analysis of residuals with SAS is not illustrated here because it was already discussed.

```
Model: MODEL1
Dependent Variable: WEIGHT              Weight in kilos
                                     Analysis of Variance
                        Sum of         Mean
     Source    DF      Squares        Square     F Value   Prob>F
     Model      1     93.32675      93.32675      31.051    0.0001
     Error     13     39.07325       3.00563                  ↑
     C Total   14    132.40000                                |
                                                          F ratio H0

        Root MSE     1.73368    R-square    0.7049
        Dep Mean    28.80000    Adj R-sq    0.6822
        C.V.         6.01971
```

Parameter Estimates

Variable	DF	Parameter Estimate	Standard Error	T for H0: Parameter = 0	Prob > |T|
INTERCEP	1	34.641720	1.13991686	30.390	0.0001 ←H0:
SIZE	1	−1.055732	0.18946059	−5.572	0.0001 $\beta = 0$

Obs.	Dep Var WEIGHT	Predict Value	Std Err Predict	Lower95% Mean	Upper95% Mean	Lower95% Predict	Upper95% Predict
1	33.0000	32.5303	0.805	30.7905	34.2700	28.4005	36.6600
2	28.0000	28.3073	0.456	27.3216	29.2931	24.4344	32.1803
3	27.0000	26.1959	0.647	24.7978	27.5939	22.1981	30.1937
4	35.0000	31.4745	0.656	30.0566	32.8924	27.4697	35.4793
5	25.0000	24.0844	0.957	22.0162	26.1526	19.8059	28.3629
6	27.0000	27.2516	0.527	26.1134	28.3898	23.3371	31.1661
7	32.0000	32.5303	0.805	30.7905	34.2700	28.4005	36.6600
8	29.0000	28.3073	0.456	27.3216	29.2931	24.4344	32.1803
9	27.0000	25.1401	0.795	23.4230	26.8573	21.0199	29.2604
10	28.0000	29.3631	0.459	28.3717	30.3544	25.4887	33.2374
11	27.0000	28.3073	0.456	27.3216	29.2931	24.4344	32.1803
12	31.0000	30.4188	0.534	29.2659	31.5716	26.5000	34.3376
13	26.0000	29.3631	0.459	28.3717	30.3544	25.4887	33.2374
14	25.0000	27.2516	0.527	26.1134	28.3898	23.3371	31.1661
15	32.0000	31.4745	0.656	30.0566	32.8924	27.4697	35.4793

Obs.	Residual	Std Err Residual	Student Residual	−2−1−0 1 2	Cook's D
1	0.4697	1.535	0.306	| | |	0.013
2	−0.3073	1.673	−0.184	| | |	0.001
3	0.8041	1.608	0.500	| | |	0.020
4	3.5255	1.605	2.197	| |**** |	0.404
5	0.9156	1.445	0.633	| |* |	0.088
6	−0.2516	1.652	−0.152	| | |	0.001
7	−0.5303	1.535	−0.345	| | |	0.016
8	0.6927	1.673	0.414	| | |	0.006
9	1.8599	1.541	1.207	| |** |	0.194
10	−1.3631	1.672	−0.815	| *| |	0.025
11	−1.3073	1.673	−0.782	| *| |	0.023
12	0.5812	1.650	0.352	| | |	0.006
13	−3.3631	1.672	−2.012	|**** | |	0.152
14	−2.2516	1.652	−1.363	| **| |	0.095
15	0.5255	1.605	0.327	| | |	0.009

Sum of Residuals   0
Sum of Squared Residuals 39.0732
Predicted Resid SS (Press) 50.7867

9.8 A research example of the use of regression

One of the aims of Hern's study (1992b) in Shipibo Indian villages in Eastern Peru was to determine the effect of modernization on fertility. A particularly interesting problem from an anthropological perspective is to determine if, as villages experience cultural change, they abandon (or nearly abandon) the practice of polygyny and if, as a result, the village's fertility declines. Therefore, the interest is to determine if fertility responds to changes in polygyny. This is an excellent case for the use of regression, in which we would want to understand the behavior of, and even predict, the fertility of a community (the dependent variable), given a measure of polygyny (the independent variable). The measure of polygyny Hern (1992b) uses is called the 'proportion of polygynous intervals', and the measure of fertility is the village's general fertility rate. The null hypothesis is that polygyny does not explain the variation of fertility, and that the slope of the regression line is not different from 0. Hern (1992b) reports a slope $b = -0.38314$ (the higher the polygyny, the lower the fertility), associated with a t score of $t = -3.873$, $p = 0.0082$. The analysis of variance yields an F ratio of $F = 14.99944$, which is significant at $\alpha = 0.0082$. The R^2 is 0.71428, and the adjusted R^2 is 0.66666. Hern (1992b) concludes that 'Polygyny is correlated with long birth intervals and negatively correlated with fertility in every respect. ...the General Fertility Rate, has the most striking and statistically significant negative association with the cumulative community index of polygyny, the proportion of polygynous birth intervals.'

9.9 Key concepts

What does a simple linear regression seek to explain?
Parametric relation
X and Y axis
Scatter shape
Linear relation
Least squares technique
Coefficient of determination
Regression equation
Residual analysis (fanning, curved behavior)
Transformation

9.10 Exercises

1. A linguist wants to determine the relationship between ability to understand mainstream culture's humor and the number of years an individual born overseas has lived in the United States. Below are data obtained in interviews from members of a single cultural minority, who were exposed to a total of 10 cartoons, and were asked to explain the reason the cartoons are considered 'funny' by mainstream Americans. What is the relationship between the variables 'Number of years lived in the United States' and 'Number of cartoons explained adequately'?

Number of years lived in the United States	Number of cartoons explained adequately
7	5
10	7
15	8
32	10
14	6
25	8
9	6
2	5
16	8
20	10
7	9
17	8
11	7
5	10
40	9
27	10
7	2
15	4
2	5
1	2
30	7
3	6
26	10
8	7
19	10
5	6
22	10

2. A researcher is interested in determining if, in a particular community, adult children's level of education can be explained by a family's income. The investigator determines the number of years of education received by the eldest son (measured after his education has ended) in a random sample of families, as well as the family's total income (rounded to the nearest 100 dollar). Use regression analysis to determine if a significant amount of education's variation can be explained by income. Follow all the steps outlined in the chapter, including residual analysis.

Family income	Number of years of education
500	14
300	10
250	7
400	8

Family income	Number of years of education
200	6
150	6
280	9
170	5
400	13
100	4
370	12
210	5
340	10
230	7
360	9
400	9
150	4

10 Correlation analysis

This chapter deals with a technique that mathematically speaking is very similar to regression analysis. Its purposes, however, are quite different. It is thus important from the start to explain what correlation analysis and its purposes are in contrast with regression.

Correlation analysis deals with two variables collected in one sample, just like regression does. However, the purpose of correlation analysis is to quantify the degree to which the variables vary together. There is no intention of explaining one variable, or of predicting one according to the other. Because correlation analysis does not deal with independent and dependent variables, our nomenclature will not distinguish between the X and Y variables, but will refer to the two as Y_1 and Y_2. To summarize then, correlation analysis simply attempts to quantify if two variables have a statistically significant co-variation. This chapter offers the reader both a parametric and a non-parametric technique for correlation analysis. The latter will be particularly useful when data can only be rank-ordered, or constitute a small sample.

10.1 The Pearson product-moment correlation

The Pearson correlation is a commonly applied technique which quantifies the relation between two variables, and tests the null hypothesis that such relation is not statistically significant. The correlation is quantified with a coefficient whose statistical symbol is 'r', and whose parametric symbol is ρ. The coefficient ranges in value from -1 to $+1$. If $r = -1$ or close to it then, as Y_1 increases, Y_2 decreases. If $r = 1$ or close to it, then as Y_1 increases, Y_2 increases as well. If $r = 0$ or is not statistically significantly different from 0, then there is no significant relation between Y_1 and Y_2. Thus, in correlation analysis the null hypothesis is that the *parametric* correlation between the two variables is 0. Thus the usual two-tail test null hypothesis is H0: $\rho = 0$. A one-tail test is possible as well, although it should be used only when there are compelling reasons for it.

The reader is by now familiar with the fact that many statistical techniques require a sample data set to be normally distributed. Indeed, for analysis of variance, it was stressed that every sample be tested for normality of distribution. Correlation analysis also has a normality assumption, one that is more difficult to test: the data need to be bivariate normal. That is, the frequency of subjects with extreme measures *in both variables* should be less frequent than the frequency of subjects with measures close to both means. Thus, a bivariate distribution looks like a three-dimensional normal curve, in which both variables have a normal distribution. The best a user with a standard computer can do is to plot the data in a flat paper, and look for obvious departures such as a curved relation or a 'fan'-like scatter (see chapter 9). If either departure is obvious, then one or both variables should be transformed. Since there are no independent and dependent variables in correlation, it does not matter which one is transformed. Thus, a plot of the data should be the first step in correlation analysis.

Mathematically speaking, the Pearson coefficient is very easily computed. In fact, the reader is familiar with all the components needed for the computation of r. Although the coefficient may be transformed into a t score, and a t table may be used for significance testing, the coefficient is more conveniently compared with a table of critical values of r, where the critical value is associated with df$= n - 2$ (table 8 in Appendix C). As usual, if the coefficient computed with the data is greater than or equal to the critical value, then the null hypothesis of no correlation is rejected. The formula for computing the Pearson correlation is:

Formula 10.1 The Pearson correlation coefficient

$$r = \frac{SP_{Y_1 Y_2}}{\sqrt{(SS_{Y_1})(SS_{Y_2})}}$$

$$\text{where } SP_{Y_1 Y_2} = \Sigma Y_1 Y_2 - \frac{(\Sigma Y_1)(\Sigma Y_2)}{n}$$

$$SS_{Y_1} \text{ and } SS_{Y_2} = \Sigma Y^2 - \frac{(\Sigma Y)^2}{n} \text{ with the appropriate subscripts}$$

Let us presume that an anthropologist is interested in finding out if there is a significant correlation between a group of females' ages at menarche, and the ages at which they experience their first birth. The anthropologist does not have a dependent variable which needs to be explained in terms of an independent one. Perhaps both variables actually co-vary, instead of one causing the other one: it is possible that females with late and early menarche commence sexual behavior at comparable ages, but that those whose menarche occurs first are able to conceive first. Or it is possible that those females with early menarche do start sexual behavior before, and as a result conceive first. The interest of the researcher is simply to establish if both variables co-vary. Thus, she tests the null hypothesis that there is no correlation between the two variables: H0: $\rho = 0$. The data are shown below:

Age at menarche	Age at first birth
13	16
12	18
11	15
11	16
12	17
12	18
13	20
13	19
11	16
12	16
12	18
11	15
12	18
13	20
13	17
14	21
12	16
12	17
11	15
10	16
13	15
12	14
13	17
13	18
12	17
14	21
13	17
11	15
12	16
14	18

$n = 30$

The first step is to plot the data, and check it for obvious departures from a bivariate normal distribution. None is observed (see figure 10.1).

Let us refer to the age at first birth as Y_1, and to the age at menarche as Y_2. We establish the following quantities:

$n = 30$, $\Sigma Y_1 = 512$, $\Sigma Y_1^2 = 8,834$, $\Sigma Y_2 = 367$, $\Sigma Y_2^2 = 4,519$ and $\Sigma Y_1 Y_2 = 6,299$

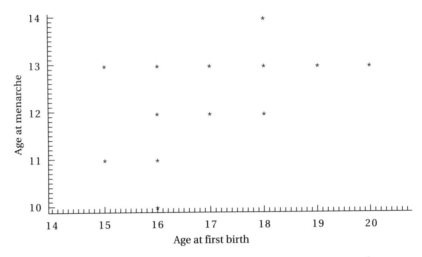

Figure 10.1. A plot of the two variables: age at first birth, and age at menarche.

First we compute the sums of products:

$$SP_{Y_1 Y_2} = 6,299 - \frac{(367)(512)}{30} = 35.53$$

The sums of squares are:

$$SS_{Y_1} = 8,834 - \frac{(512)^2}{30} = 95.87$$

$$SS_{Y_2} = 4,519 - \frac{(367)^2}{30} = 29.37$$

Therefore:

$$r = \frac{35.53}{\sqrt{(95.87)(29.37)}} = 0.67$$

With degrees of freedom $df = 28$, our correlation coefficient is significant at the 0.01 level. Our conclusion is that both variables are highly correlated in this community.

Practice problem 10.1

An anthropologist suspects that, in the community in which he works, there is a correlation between the number of domestic goats a female owns, and the number of marriages she has had. The researcher interviews females who are at least 50 years of age, and finds out how many marriages the female has had, and how many goats she owns. The data are below:

Number of marriages at time of interview	Number of goats at time of interview
4	7
3	8
0	15
1	3
1	9
2	10
1	9
2	6
1	11
1	5
3	10
0	9
2	12
3	8
1	6
2	11
1	12
2	6
4	13
3	7
0	10
1	8
2	9

$n = 23$

The first step is to look for obvious departure from bivariate normality:

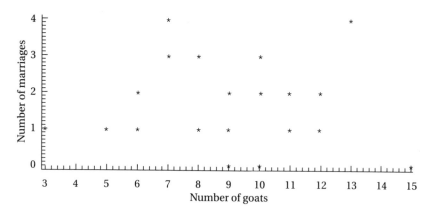

The data do not appear to suffer from curved or 'fan'-like behavior.

Let us refer to the number of goats as Y_1 and to the number of marriages as Y_2. We then establish the following statistics:

$$n = 23, \Sigma Y_1 = 204, \Sigma Y_1^2 = 1{,}980, \Sigma Y_2 = 40, \Sigma Y_2^2 = 100, \text{ and } \Sigma Y_1 Y_2 = 350$$

First we compute the sums of products:

$$SP_{Y_1 Y_2} = 350 - \frac{(204)(40)}{23} = -4.78$$

The sums of squares are:

$$SS_{Y_1} = 1{,}980 - \frac{(204)^2}{23} = 170.61$$

$$SS_{Y_2} = 100 - \frac{(40)^2}{23} = 30.43$$

Therefore:

$$r = \frac{-4.78}{\sqrt{(170.61)(30.43)}} = -0.067$$

With degrees of freedom $df = 21$, our correlation coefficient is clearly not significantly different from 0 $(cv = 0.413)$.

The Pearson correlation using SAS/ASSIST

The data need to be entered into two columns, each one of which has a variable. The path to obtain the graph is: graphics, high resolution, and simple $X*Y$ plot. At this window, the user specifies the names of the variables to be plotted. To do the correlation, the path is: data analysis, elementary, correlation. At this window, the user specifies the form of correlation desired. Below is the output obtained for the latter example (goats and number of marriages).

SAS prints the correlation results in the form of a matrix, that is: the correlation between goats and marriage is the same as the correlation between marriage and goats, which is -0.067 with a p value of 0.76. The correlation between marriage and itself, and between goats and itself, is 1, with a p value 0.

```
                    Correlation Analysis
                2 'VAR' Variables: MARRIAGE GOATS

    Pearson Correlation Coefficients / Prob >|R|under Ho: Rho=0 / N=23
                                        MARRIAGE        GOATS
    MARRIAGE                            1.00000    -0.06637   ←r
    Number of marriages at time of interview  0.0    0.7635   ←p
    GOATS                              -0.06637     1.00000
    Number of goats at time of interview      0.7635      0.0
```

10.2 A research example of the use of Pearson correlation

In their paper entitled 'The health and nutrition of a medieval Nubian Population', Van Gerven, Sheridan and Adams (1995) investigate the skeletal populations from two cemeteries at Kulubnarti, Nubia. One of their main focuses of interest is to understand the pattern of childhood stress and disease in 'the population's most important component, its children' (Van Gerven et al., 1995: 477). Two well-established indicators of childhood stress are present in the population: Cribra orbitalia (a porotic lesion in the upper orbit), and enamel hypoplasia (bands in the enamel which indicate growth interruption). The researchers were interested in determining if there is a significant correlation between both variables, that is, if they occur as part of an overall childhood stress syndrome. Van Gerven et al. report that among children less than 5 years of age there is a highly significant correlation between both variables: $r = 0.98$, $p < 0.01$ (Van Gerven et al., 1995: 474).

10.3 The Spearman correlation

There are two well-known non-parametric tests which test the null hypothesis that there is no association between two variables: The Kendall's coefficient of rank, and Spearman's coefficient r_s. The former is not covered here because it involves some rather elaborate hand calculations and requires a special table for $n < 40$. The Spearman's coefficient rests on the same ranking principles we already covered in the non-parametric chapter, and has the added advantage that, for sample sizes greater than 10, the table of critical values for the Pearson correlation may be used. Moreover, the Spearman's coefficient is more appropriate than is Kendall's if the researcher is not totally certain about the rank-order position of several variates (for example, if two individuals are ranked equally or virtually equally in a prestige scale). Finally, r_s ranges in value from -1 to $+1$, just like Pearson's.

The Spearman correlation coefficient requires that the relation between the variables be linear. Therefore, the data must still be plotted and inspected. The data may be numerical data which are approximate, or even two columns of ranks. Therefore, this test can be applied to a wide variety of research interests.

The test proceeds as follows (after the inspection of the plot).
1. Rank both variables, assigning ties if necessary. The rest of the test works with the two columns of ranks. Frequently the data will be the ranks themselves.
2. Rank-order one of the variables, and assign the other rank to the appropriate observation. Therefore, one of the columns will be ranked. The other may or may not be. The null hypothesis tests that there is no association

between the two columns of ranks. If there is total agreement (that is, if the same observation is ranked in the same level in both columns), then the null hypothesis will be rejected. If there is no agreement, the null hypothesis of no association will be accepted.

3. Obtain the difference (*D*) between the two ranks and square it (*D²*). The formula for the r_s coefficient is:

Formula 10.2 The Spearman rank-order correlation coefficient

$$r_s = 1 - \left[\frac{(6)(\Sigma D^2)}{(n)(n^2 - 1)} \right], \quad df = n - 2$$

where *D* is the difference between the two columns of ranks.

Let us work first with a data set that includes both a quantitative measure and a set of ranks as its two variables. Let us presume that an anthropologist is interested in investigating if, in a particular culture, prestige among adult males is correlated with the number of children the males produce. The researcher first interviews a member of the community who is said to be outside the prestige hierarchy of adult males: the religious specialist. The anthropologist asks the informant to rank-order the adult-male heads of households in the community. This variable consists of ranks only, where the highest ranked male receives the highest number (11). The researcher then proceeds to interview each male, and asks him how many children he has produced. Thus, the number of children reported, though likely to be an approximation, is a quantitative measure. Below are the data:

Individual interviewed (ID)	Prestige rank	Number of children	Number of children (ranked)
1	11	15	11
2	9	12	9.5
3	1	3	2
4	2	0	1
5	10	12	9.5
6	3	5	3
7	8	7	5
8	5	10	7.5
9	7	7	5
10	6	7	5
11	4	10	7.5

The first step is to plot the data and determine if there is a violation to the assumption of a linear relation between the variables (see figure 10.2). The data do appear to be linearly related.

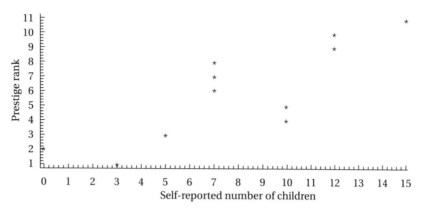

Figure 10.2.

We then rank-order the data according to one of the columns. Since prestige is already ranked, we can order the observations according to it. The number of children was ranked above, assigning ties when necessary. The ranked data are shown below, as are the differences between the two ranks:

Obs.	ID	Prestige	Children	D	D^2
1	3	1	2.0	-1	1
2	4	2	1.0	1	1
3	6	3	3.0	0	0
4	11	4	7.5	-3.5	12.25
5	8	5	7.5	-2.5	6.25
6	10	6	5.0	1	1
7	9	7	5.0	2	4
8	7	8	5.0	3	9
9	2	9	9.5	-0.5	0.25
10	5	10	9.5	0.5	0.25
11	1	11	11.0	0	0
				$\Sigma D^2 =$	35

$$r_s = 1 - \left[\frac{(6)(35)}{(11)(11^2 - 1)} \right] 1 - 0.16 = 0.84, \quad df = 9, \quad p < 0.01$$

We reject the null hypothesis of no association between prestige and number of children with a high degree of confidence.

Practice problem 10.2

An archaeologist is interested in evaluating the consistency of opinions held by different archaeologists about archaeological societies' reliance on agriculture. The researcher asks two experts to rank 12 archaeological cultures on their reliance on agriculture, with the most reliant society receiving the highest rank. The null hypothesis tested is that there is no association between the archaeologists' ranks. Notice that, in this example, the data themselves are ranks. Below are the data and their plot:

Obs.	Society	Rank 1	Rank 2
1	1	3	1
2	2	1	3
3	3	9	5
4	4	12	10
5	5	5	2
6	6	11	11
7	7	2	4
8	8	6	8
9	9	8	12
10	10	10	7
11	11	7	6
12	12	4	9

Two independent rankings of 12 archaeological societies' dependence on agriculture.

No violation of the assumption of linear relation is evident.

The data are first ranked according to the variable Rank 1, and the difference between both ranks is computed:

Obs.	Society	Rank 1	Rank 2	D
1	2	1	3	-2
2	7	2	4	-2
3	1	3	1	2
4	12	4	9	-5
5	5	5	2	3
6	8	6	8	-2
7	11	7	6	1
8	9	8	12	-4
9	3	9	5	4
10	10	10	7	3
11	6	11	11	0
12	4	12	10	2

$$\Sigma D^2 = 96$$

$$r_s = 1 - \left[\frac{(6)(96)}{(12)(12^2 - 1)} \right] 1 - 0.33 = 0.67, \quad df = 10, \quad p < 0.02$$

We reject the null hypothesis of no association at the 0.02 α level. There is a significant relation between the two experts' rankings.

Spearman correlation using SAS/ASSIST

The path followed does not differ from that already described in the previous section (for Pearson correlation). The only difference is that at the correlation window, the user clicks on the Spearman, instead of the Pearson correlation. Below is the output of the latter practice (the two archaeologists' rankings). The output is presented in the same manner as that of the Pearson correlation:

```
                        Correlation Analysis
                   2 'VAR' Variables: RANK1 RANK2

   Spearman Correlation Coefficients / Prob >|R| under H0: Rho = 0 / N = 12
                                        RANK1      RANK2

RANK1                                 1.00000    0.66434   ←r_s
Rank given by first archaeologist        0.0     0.0185   ←p
RANK2                                 0.66434    1.00000
Rank given by second archaeologist    0.0185        0.0
```

10.4 A research example of the Spearman correlation coefficient

Pollitzer, Smith and Williams (1988) investigate inbreeding and migration in a Yorkshire Parish from 1654 through 1916 by studying isonymic marriages. The study of isonymy (marriage of individuals with the same last name) provides an useful tool to the understanding of gene flow and inbreeding in a community, since surnames are inherited like the Y chromosome. Pollitzer et al. (1988) compute a number of inbreeding coefficients (among others, random, non-random, and total inbreeding), and correlate them among themselves and with the number of marriages. The frequency of isonymous marriages is significantly correlated with the index of non-random inbreeding ($p < 0.001$, although the actual value of the correlation coefficient is not reported). The authors conclude that the non-random inbreeding 'varies widely through the decades, tending to reflect the number of isonymous marriages' (Pollitzer et al., 1988). This is an excellent example of a data set which is better analyzed with a non-parametric test, since the data were rates and coefficients, instead of actual quantitative measurements.

10.5 Key concepts

The difference between correlation and regression
Pearson correlation: what does it quantify?
Bivariate normal distribution
When is the Spearman correlation appropriate?

10.6 Exercises

1. A social anthropologist is working with female victims from a particular suburban region who are recovering from domestic (spousal) abuse. She interviews several women and asks them to estimate how many instances of physical abuse (e.g. hitting, shoving) they endured during the last year of marriage. She then asks what the male spouse's income was for that year. Is there a correlation between instances of physical abuse and income?

Woman	Instances of physical abuse	Male spouse's income ($)
1	360	36000
2	70	42000
3	50	25000
4	200	40000
−5	100	48000
7	80	25000
8	150	75000
9	300	100000
10	90	50000
11	200	0
12	400	60000
13	150	45000
14	365	150000
15	500	28000
16	200	190000
17	100	30000
18	90	26000
19	50	0
20	300	58000
21	250	200000
22	365	60000
23	500	45000
24	150	28000
25	400	32000
26	350	70000
27	500	25000
28	200	20000
29	360	36000
30	250	50000

2. A medical anthropologist wants to determine if there is a significant correlation between age and how well one scores on an exam which tests cognitive functions. The test, which has a total possible score of 100, was given to an elderly cohort whose ages range from 67 to 88 ($n=25$). Compute the Spearman correlation coefficient to determine if there is a significant correlation between age and test score.

Age	Score	Age	Score
68	95	79	90
78	77	84	68
82	60	70	67
68	98	87	96
75	100	81	93
80	66	73	75
85	97	76	79
77	82	75	60
67	88	84	70
73	89	71	77
88	76	85	75
86	79	86	70
69	90		

11 | The analysis of frequencies

This chapter presents a departure from the statistical tests covered so far: we do not deal with quantitative variables such as fertility, prestige ranks, height, weight, etc., within a sample or among several samples. Here we deal with frequencies of occurrences or events. We work with one variable which has a few possible outcomes, and are concerned with the number of individuals who have each of the outcomes (where a is the number of outcomes). Often-used anthropological frequency data are gene frequencies, the frequencies of males and females, the frequencies of different ethnic groups, the frequencies of different pottery styles in archaeological sites, etc.

This chapter covers the well-known and widely used chi-square test (X^2), applied as a goodness-of-fit test and as a test for independence of two variables. Not only is the X^2 test widely used, but unfortunately it has also been widely misused. Therefore, a correction which must be applied to small data sets is also discussed. The most common error about the X^2 test is its own name: it is not infrequently referred to as χ^2, which is the parametric notation for a theoretical frequency distribution against which the statistic computed from the sample (X^2) should be compared.

11.1 The X^2 test for goodness-of-fit

The purpose of this test is to determine if the observed frequencies of events depart significantly from frequencies proposed by a null hypothesis. How that null hypothesis is generated, depends on the specific research project. Usually, the expected frequencies (f_e) are generated in either of two ways: the frequencies of observations in all outcomes are expected to be the same, that is, there is an equal probability associated with the various outcomes; or the frequencies are generated by an expectation based on knowledge of the data's nature. An example of the first case would be to compare the number of observed males and females in (e.g.) medical school, with an expected distribution that half the students would be males, and half would be females (if we sample $n = 70$ students, then the expected frequencies of male students would be $f_{males} = 70/2 = 35$, and the expected frequencies of female students would be $f_{females} = 70/2 = 35$). An example of the second null hypothesis generation is when we compare observed gene frequencies with those expected under the assumptions of the Hardy–Weinberg equilibrium. That is, if the population is not evolving and mates randomly, then we would expect to see specific frequencies of phenotypes. We use a X^2 to compare such expected frequencies with the observed ones. However the expected frequencies of the null hypothesis are generated, the mathematical calculations for the computation of the X^2 test are the same. The test is not parametric, since it does not make statements about the parameters or distribution of the population from which the sample(s) was (were) obtained. All the null hypothesis states is that H0: $f_o = f_e$, that is, that the observed frequencies are not significantly different from those expected if the null hypothesis is true. Although non-parametric, the X^2 still assumes random sampling and that the expected frequencies be at least 5. The first assumption is a matter of the research design. The second assumption must be attended to: if it is violated, the computation of X^2 must be corrected. This will be covered later in the chapter. The chi-square test proceeds in the following fashion.

1. State the null hypothesis, and generate the expected frequencies. Usually the researcher states how the frequencies were obtained, if by an expectation of equality among all outcomes, or by an expectation based on previous studies or knowledge (like the Hardy–Weinberg equilibrium).

2. Arrange the data so that the observed and expected frequencies are in the same cell. For example, if we collected a sample of 70 medical school students, and want to test the null hypothesis that the frequency of males is the same as the frequency of females, that is H0: $f_{females} = f_{males} = 70/2 = 35$, and observe 30 females and 40 males, then we would arrange the data as follows:

	Females	Males
Observed (f_o)	30	40
Expected (f_e)	35	35

3. Compute the X^2 statistic as follows:

....

Formula 11.1 Formula for the computation of the X^2

$$X^2 = \Sigma \frac{(f_o - f_e)_9}{f_e}, \text{ across all } a \text{ outcomes, with}$$

$$df = a - 1.$$

....

4. If our X^2 is equal to or greater than the critical value, then we reject the null hypothesis.

We first practice with the data presented above on the gender of a sample of 70 medical students. The X^2 is:

$$X^2 = \Sigma \frac{(30 - 35)_9}{35} + \frac{(40 - 35)_9}{35} = 0.7 + 0.71 = 1.43$$

Since there are two categories (males and females) then df$=2-1=1$. Since the critical value at 0.05 is 3.841, we accept the null hypothesis that the observed frequencies of male and female students do not differ from those expected under the null hypothesis.

Practice problem 11.1

We now practice the goodness-of-fit test with an example in which the null hypothesis is generated according to a previous study. Earlier in this book, we computed the frequencies of ceramic types from test unit A in Depot Creek shell mound (data from White, 1994). We can use these frequencies to generate the null hypothesis for another test unit. That is, we could test the hypothesis that there is no difference between the observed frequencies (test unit B) and the expected frequencies (generated from test unit A). The data from test unit A are:

Ceramic type	f	p
Sand-tempered	45	0.141
Grog-tempered	9	0.028
Check-stamped	184	0.58
Indent-stamped	28	0.088
Simple-stamped	43	0.135
Other	9	0.028
	$n=318$	$\Sigma p=1$

Note that two categories (comp-stamped and cord-stamped) have been merged into one ('other') because their individual contribution to the total sample size was small. If they had been left separately, the assumption that the minimum expected frequency per category be 5 would have been violated. With these data from test unit A, we can generate expected frequencies for test unit B. Since 14.1% of the ceramics in test unit A are sand-tempered, we expect 14.1% of the ceramics in test unit B to be sand-tempered. Thus, we simply multiply the sample size from B ($n=216$) by the expected frequency: $(216)(0.141) = 30.46$. We compute the other expected frequencies in the same manner:

Ceramic types from Depot Creek shell mound (86u56), test unit B

Ceramic type	f_o	f_e		f_o-f_e	$\dfrac{(f_o-f_e)^2}{f_e}$
Sand-tempered	40	$(216)(0.141) =$	30.46	9.54	2.99
Grog-tempered	15	$(216)(0.028) =$	6.05	8.95	13.24
Check-stamped	88	$(216)(0.58) =$	125.28	-37.28	11.1
Indent-stamped	46	$(216)(0.088) =$	19.00	27.0	38.37
Simple-stamped	15	$(216)(0.135) =$	29.16	-14.16	6.87
Other	12	$(216)(0.028) =$	6.05	5.95	5.85
	$n=216$			$\Sigma=0$	$\Sigma=78.42$

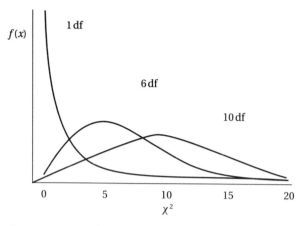

Figure 11.1. The χ^2 distribution for several degrees of freedom.

Since there are six categories, then df = 5. We reject the null hypothesis, because our X^2 is greater than the critical value at 0.05 and even at 0.001. We conclude that the two test units differ in their frequencies of ceramic types.

Before we continue, the reasoning behind the computation of the X^2 formula should be discussed, specifically, why we square the deviation between the observed and expected frequencies and divide it by the expected frequencies. Notice that the X^2 statistic computes the 'raw' difference between the observed and expected frequencies. The difference must then be squared so that the positive and negative differences do not cancel each other out (indeed, if the 'raw' differences from the previous example are added, they sum to 0, as shown above). The reason these squared deviations ought to be divided by the expected frequency is that, by dividing the difference, it is expressed as a proportion of the expected frequency. Therefore, if the squared deviation is virtually equal to the expected frequency then we would be dividing a number basically by itself, so the quotient of the division would be close to 1. In this situation, we are not likely to reject the null hypothesis. Now, if the squared deviation is a large number because there is a large difference between f_o and f_e, then we would be dividing a large number by a smaller one, the quotient of which division would be greater than 1. In this situation, we would be likely to reject the null hypothesis. Finally, please note that, since we work with square deviations, the χ^2 distribution is not symmetrical but cuts off at 0. Figure 11.1 shows the χ^2 distribution for several degrees of freedom.

11.2 A research example of the X^2 test for goodness-of-fit

In their study on health changes at Dickson mounds during AD 950–1300, Goodman et al. (1984) investigate if physiological stress may be related to the site's increasing involvement in Mississippian-based exchange systems. The site has three cultural horizons, namely Late Woodland (LW, circa AD 950–1100), Mississippian Acculturated Late Woodland (MALW, AD 1100–1200) and Middle Mississippian (MM, AD 1200–1300). Goodman et al. (1984) use X^2 to test the hypothesis that the frequency of cases with one or more growth disruptions (namely enamel hypoplasias) and the frequency of cases with no growth disruptions did not differ by cultural level. They report that 'the hypoplasia data support the hypothesis that physiological disruption was more frequent and severe in the Middle Mississippian' (Goodman et al., 1984: 286).

11.3 The X^2 test for independence of variables

The X^2 test for independence of variables does not differ from the goodness-of-fit test in its null hypothesis or even computation (although an easier computation formula may be applied). The purpose is still to determine if the observed frequency of observations falling into specific outcomes is significantly different from those proposed by the null hypothesis, however the latter may have been generated. However, the outcomes are now determined by two variables, each of which has at least two possible outcomes. Therefore, the null hypothesis may now be re-worded as H0: the two variables are independent. Thus, we could test if the number of male and female medical students is independent of socio-economic status (defined as a discrete variable, say high, low, and middle). Or, we could test that the frequency of six ceramic types is independent from the stratigraphic level from which they were obtained. Or, the frequency of several hemoglobin types may be tested across two different ethnic groups, etc. The data are usually arranged in a table with r rows and c columns, where the number of individuals that belong to each cell (defined by the two categories) is displayed. For example, the number of individuals with the first outcome in both categories is displayed in cell a, the number of individuals with the first outcome in the first category and the second outcome in the second category is displayed in cell d, etc. The table below is a 3×3 table, since it has three rows and three columns (it is customary to refer to the number of rows first, and to the number of columns second). Notice that the column and row totals are additive, and that the grand total is equal to the sample size:

		First variable			
		First outcome	*Second outcome*	*Third outcome*	**Row totals**
Second variable	*First outcome*	$f_o = a$	$f_o = b$	$f_o = c$	RT1 $= a + b + c$
	Second outcome	$f_o = d$	$f_o = e$	$f_o = f$	RT2 $= d + e + f$
	Third outcome	$f_o = g$	$f_o = h$	$f_o = i$	RT3 $= g + h + i$
Column totals		CT1 $= a + d + g$	CT2 $= b + e + h$	CT3 $= c + f + i$	Grand total $= n$

The H0's expected frequencies can be generated by a previous study, although they are usually generated as if the two variables are independent of each other. This is done by multiplying a cell's column (column total or CT) and row totals (or RT), and dividing by the grand total. Care must be taken that the expected frequencies be additive. Thus, the expected frequencies are computed as follows:

		First variable			Row totals
		First outcome	*Second outcome*	*Third outcome*	
Second variable	*First outcome*	$[(CT1)(RT1)]/n$	$[(CT2)(RT1)]/n$	$[(CT3)(RT1)]/n$	**RT1**
	Second outcome	$[(CT1)(RT2)]/n$	$[(CT2)(RT2)]/n$	$[(CT3)(RT2)]/n$	**RT2**
	Third outcome	$[(CT1)(RT3)]/n$	$[(CT2)(RT3)]/n$	$[(CT3)(RT3)]/n$	**RT3**
Column totals		**CT1**	**CT2**	**CT3**	**Grand total** $= n$

The formula for the computation for the X^2 statistic remains the same: we square the difference between the observed and expected frequencies, divide by the expected value, and sum across all cells. The formula for the degrees of freedom takes into consideration the number of columns and rows: $\mathrm{df} = (c-1)(r-1)$.

Let us presume that an anthropologist is interested in determining if, in a particular culture, female property ownership is related to the female's marital status. The researcher collects data about the adult (as defined in the society) females, consisting of the number of subjects who are married and unmarried, and who do and do not own property. The data are below:

		First variable: marital status		
		Married	Unmarried	
Second variable:	Own property	40	6	$40 + 6 = 46$
property ownership	Do not own property	20	10	$20 + 10 = 30$
		$40 + 20 = 60$	$6 + 10 = 16$	$n = 60 + 16 = 76$
				or $n = 46 + 30 = 76$

The first step is to compute the expected frequencies as follows:

		First variable: marital status		
		Married	Unmarried	
Property	Own property	$[(46)(60)]/76 = 36.32$	$[(46)(16)]/76 = 9.68$	46
ownership	Do not own property	$[(30)(60)]/76 = 23.68$	$[(30)(16)]/76 = 6.32$	30
		60	16	$n = 76$

Note that, since the expected frequencies are additive, only some of them needed to be computed in this manner. For example, for the first column, after we computed the f_e for cell a (36.32), we could have obtained the f_e for cell c as $60 - 36.32 = 23.68$. Having computed the expected frequencies for the first column, we can obtain the frequencies for the second column by subtracting from the row totals the expected frequencies of the first column. Thus, the expected frequency for unmarried property-owners is:

$46 - 36.32 = 9.68$. The expected frequency of unmarried non-property-owners is $30 - 23.68 = 6.32$. We now compute our X^2 statistic:

$$X^2 = \frac{(40-36.32)^2}{36.32} + \frac{(6-9.68)^2}{9.68} + \frac{(20-23.68)^2}{23.68} + \frac{(10-6.32)^2}{6.32} = 4.49$$

The degrees of freedom are $(2-1)(2-1) = 1$. Since the critical value at $=0.05$ is 3.841, we reject the null hypothesis that adult female property ownership is independent of marital status in this community: there is a significant association between both variables.

Practice problem 11.2

A medical anthropologist is interested in investigating if in the community she is studying, mothers who use traditional medicine to heal their children are significantly more or less educated than mothers who do not. The anthropologist divides the level of education into three levels: low, medium and high. Test the null hypothesis that use of traditional medicine is unrelated to educational level. Below are the data:

Education level of mothers	Use of traditional medicine
low	no
medium	no
high	no
low	no
low	no
low	no
low	no
low	no
low	no
low	no
medium	no
medium	no
medium	no
high	no
high	no
high	no
high	no
high	no
high	no
high	no
high	no
high	no
high	no

Education level of mothers	Use of traditional medicine
high	no
high	no
high	no
low	yes
medium	yes
high	yes
low	yes
low	yes
low	yes
low	yes
low	yes
low	yes
low	yes
low	yes
low	yes
low	yes
low	yes
low	yes
low	yes
medium	yes
medium	yes
medium	yes
medium	yes
high	yes
high	yes

$$n = 48$$

The first step is to arrange the data into a table that includes the observed and the expected frequencies. The expected frequencies are written in bold face. Notice that they add to the column and row totals:

```
Table of 'use' by 'education'
Use (Use of traditional medicine)
              Education (Education level of mothers)
```

Frequency	high	low	medium	Total
no	14 **9.21**	8 **11.92**	4 **4.87**	26 54.17
yes	3 **7.79**	14 **10.08**	5 **4.13**	22 45.83
Total	17 35.42	22 45.83	9 18.75	48 100.00

$$X^2 = \frac{(14-9.21)9}{9.21} + \frac{(8-11.92)9}{11.92} + \frac{(4-4.87)9}{4.87} + \frac{(3-7.79)9}{7.79} + \frac{(14-10.08)9}{10.08} + \frac{(5-4.13)9}{4.13} = 8.59, \text{df}$$

$$= (3-1)(2-1) = 2$$

The critical values at $\alpha = 0.025$ and $\alpha = 0.01$ are 7.378 and 9.210 respectively. Thus, we reject the critical value at an alpha level less than 0.025 but greater than 0.01. Our data indicate that educated females use less and non-educated females use more traditional medicine than expected, and that this difference is significant.

11.4　A research example of the X^2 test for independence of variables

In their article titled 'Salience counts: a domain analysis of English color terms', Smith et al. (1995) analyze with a salience index the frequency and rank of order of mention of English color terms. A higher value of salience indicates a higher frequency and a higher rank of order of mention. Smith et al. (1995) divide color terms into five categories (simple basic, simplex secondary opaque, simplex secondary transparent, complex, and residual), and report that the five terms formed distinct clusters within the salience distribution. In this distribution, the terms with higher salience are placed in the upper half, and those with lower salience in the lower half. The authors used a 2×2 chi-square test to test the hypothesis that the secondary opaque and transparent terms in the upper and lower half of the distribution are distributed equally. Thus, the null hypothesis states that the placement in the salience distribution (upper or lower) is independent from color groups (here two categories: secondary opaque and transparent terms). They report that 'Some 90 percent of the secondary opaque terms are in the upper half, while less than 60 percent of the transparent terms fall there' (Smith et al., 1995: 211). The results are significant with a X^2 value of 9.586, df = 1, and $p < 0.01$. Thus, these two categories of color terms have a significantly different salience distribution.

11.5　Yates' correction for continuity

It was mentioned previously that the X^2 test assumes that the cell's expected frequencies be at least 5. In cases in which this assumption does not hold, the X^2 statistic should be corrected. However, there is some debate about which correction is best, as well as about how closely this assumption must be followed. The correction of choice for a 2×2 table with small sample sizes is Yates' correction for continuity. The main advantage of Yates' correction is its easy calculation, while its main disadvantage is that it is too conservative for tables larger than a 2×2. For such tables, other corrections, more cumbersome and involved mathematically, are recommended. This book covers Yates' because anthropologists frequently encounter small sample sizes, and this is the correction of choice in such cases. Moreover, it is very easy to compute by hand. If the user has access to a computer, then one of the other corrections (Fisher's particularly) can easily be obtained.

The formula for Yates' correction involves a minor change from the usual X^2 computation: after the difference between the f_o and the f_e is obtained, 0.5 is extracted from the difference's absolute value. That is, if the difference were 2, then we would correct it as $2 - 0.5 = 1.5$. The reader can see that Yates' correction diminishes the difference between the observed and expected frequencies, making the test more conservative. To distinguish a corrected from a non-corrected X^2, the former is usually denoted as X_c^2. The formula for Yates' correction is:

Formula 11.2 Yates' correction for continuity

$$X_c^2 = \Sigma \frac{(|f_o - f_e| - 0.5)^9}{f_e}$$

with summation across the a categories, and df as before

Let us practice the correction with the data set concerning the mothers' education and their use of traditional medicine, since there were two cells whose expected frequency was under 5: 4.87 and 4.13. The X^2 is computed as follows:

$$X_c^2 = \frac{(|14 - 9.21| - 0.5)^9}{9.21} + \frac{(|8 - 11.92| - 0.5)^9}{11.92} + \frac{(|4 - 4.87| - 0.5)^9}{4.87}$$
$$+ \frac{(|3 - 7.79| - 0.5)^9}{7.79} + \frac{(|14 - 10.08| - 0.5)^9}{10.08} + \frac{(|5 - 4.13| - 0.5)^9}{4.13}$$
$$= 1.99 + 0.98 + 0.03 + 2.4 + 1.16 + 0.033 = 6.6$$

$$\text{df} = (3 - 1)(2 - 1) = 2.$$

We reject the null hypothesis at the level of 0.05, but not at the level of 0.025. Our X^2 has diminished by almost two units. Clearly, Yates' correction is more likely to lead to the retention of the null hypothesis.

Chi-square using SAS/ASSIST

For a goodness-of-fit test in which the null hypothesis proposes all outcomes should have the same frequency, the data are entered into a single column. If the null hypothesis states specific frequencies (as when we compared the observed frequencies in a test unit with those obtained in another test unit in the same archaeological site), the user is better off doing the test by hand. For an independence-of-variables test, the data are entered as follows (only part of the data set is shown):

Education level of mothers	Use of traditional medicine
low	no
medium	no
high	no
low	no

The user has several options for requesting contingency tables which include the percentages, row and column totals, etc. Three such paths are: report writing, counts, and cross-tabulation. Another option is to go to the index, and click on frequency (tables), analysis of existing table. At this point the user enters the number of observations per cell, and is able to obtain X^2 statistics. Lastly, the user can also follow this path: data analysis, elementary, frequency tables. These various options yield comparable outputs, such as:

```
                  Table of 'use' by 'education'
                  Use              Education
        Items in  Frequency
        cells:    Percent
                  Row Pct
                  Col Pct        1         2         3         Total

                        1        14        8         4         26
                                 29.17     16.67     8.33      54.17
                                 53.85     30.77     15.38
                                 82.35     36.36     44.44

                        2        3         14        5         22
                                 6.25      29.17     10.42     45.83
                                 13.64     63.64     22.73
                                 17.65     63.64     55.56

                  Total          17        22        9         48
                                 35.42     45.83     18.75     100.00
```

```
Statistics for table of use by education
Statistic                      df    Value         Prob
─────────────────────────────────────────────────────────
Chi-square                     2     8.591    0.014←uncorrected
Fisher's Exact Test (2-Tail)                  0.012←corrected
Sample Size = 48                                      statistic
WARNING: 33% of the cells have expected counts less
         than 5. Chi-square may not be a valid test.
```

Notice that SAS warns you about the expected frequencies being less than 5. In such case, use the Fisher's exact test.

11.6 Key concepts

What are the null hypothesis of chi-square (for both goodness-of-fit and independence-of-variable tests)

What is the correct notation for chi-square?

Why is the X^2 not parametric?

Refer to the computational formula of X^2. Why is the deviation between the observed and expected frequencies computed the way it is?

Yates' correction for continuity

11.7 Exercises

1. An anthropologist working with a horticultural group wishes to determine if the group's meat consumption differs seasonally. The researcher records the number of large animals consumed (shared by) the entire community during the rainy and dry seasons. Test the hypothesis that the frequency of animals consumed does not differ seasonally.

```
Table of season by consumed
Season (Seasonality divide into wet and dry)
            Consumed (Large animal reported consumed)
```

Frequency %	Large	Total
Dry	40	40
	36.36	36.36
wet	70	70
	63.64	63.64
Total	110	110
	100.00	100.00

2. An anthropologist working with a horticultural group notices that when the community shares the meat of a large animal, the hunter gives as a present the head of the animal to somebody of his choice. Both males and females of high and low status receive the present. The anthropologist records the following data set. Use it to test the hypothesis that gender and status are independent.

```
Table of status by gender
Status (Two levels: high and low)
          Gender
```

Frequency % Row % Col %	female	male	Total
high	17	20	37
	26.56	31.25	57.81
	45.95	54.05	
	48.57	68.97	
low	18	9	27
	28.13	14.06	42.19
	66.67	33.33	
	51.43	31.03	
Total	35	29	64
	54.69	45.31	100.00

REFERENCES

Baker, T. L. 1994. *Doing social research.* McGraw-Hill. Second edition: New York

Blakeslee, A. F. 1914. Corn and men. *The Journal of Heredity.* 5: 511–18

Bohrnstedt, G. W. and D. Knoke. 1988. *Statistics for social data analysis,* second edition. F. E. Peacock Publishers, Inc.: Itasca

Bramblett, C. A. 1994. *Patterns of primate behavior,* second edition. Waveland Press, Inc. : Prospect Heights

Close, F. E. 1991. *Too hot to handle: the race for cold fusion.* Princeton University Press: Princeton

Draper, N. R. and H. Smith. 1981. *Applied regression analysis,* second edition. John Wiley and sons: New York

Ellis, L. 1994. *Research methods in the social sciences.* WCB Rown and Benchmark publisher: Dubuque

Feldesman, M. R. and R. L. Fountain. 1996. 'Race' specificity and the femur/stature ratio. *American Journal of Physical Anthropology* 100: 207–24

Fisher, R. A. 1959. *Statistical methods and scientific inference.* Hafner Publishing Company: New York

Fisher, R. A. 1991a. *Statistical methods for research workers* (reprinted with corrections from original volumes). Oxford University Press: Oxford

Fisher, R. A. 1991b. *The design of experiments* (reprinted with corrections from original volumes). Oxford University Press: Oxford

Fleiss, J. L. 1986. *The design and analysis of clinical experiments.* John Wiley: New York

Futuyma, D. J. 1995. *Science on trial: The case for evolution.* Sinauer Associates, Inc: Sunderland

Goodman A. H., J. Lallo, G. J. Armelagos and J. C. Rose. 1984. Health changes at Dickson Mounds, Illinois (A.D. 950–1300). In: *Paleopathology at the origins of agricuture.* eds. M. N. Cohen and G. J. Armelagos. Academic Press: Orlando pp. 271–305

Gould, S. J. 1990. An earful of jaw. *Natural History* 3: 12–23

Gould, S. J. 1993. Cordelia's dilemma. *Natural History* 2: 10–18

Gravetter, F. J. and L. B. Wallnau. 1992. *Statistics for the behavioral sciences,* third edition. West Publishing company: St Paul

Gray, S. J. 1996. Ecology of weaning among nomadic Turkana pastoralists of Kenya: maternal thinking, maternal behavior, and human adaptive strategies. *Human Biology* 68 (3): 437–65

Hern, W. M. 1992a. Polygyny and fertility among the Shipibo of the Peruvian Amazon. *Population Studies,* 46: 53–64

Hern, W. M. 1992b. Shipibo polygyny and patrilocality. *American Ethnologist* 19 (3): 501–22

Jenkins, G. M. 1979. *Practical experiences with modelling and forecasting time series.* Time Series library, a GJP publication: Jersey, Channel Islands

Judd, C. M., E. R. Smith and L. H. Kidder. 1991. *Research methods in social relations,* sixth edition. Holt, Rinehart and Winston Inc: New York

Klockars, L. J. and G. Sax. 1986. Multiple comparisons. A Sage University paper. Series: Quantitative applications in the social sciences #61

Kramer, A. 1993. Human taxonomic diversity in the Pleistocene: does Homo erectus represent multiple hominid species? *American Journal of Physical Anthropology* 91: 161–71

Leatherman, T. 1994. Health implications of changing agrarian economies in the Southern Andes. *Human Organization* 53(4): 371–80

Lisker R., E. Ramirez and V. Babinsky. 1996. Genetic structure of autochtonous populations of Meso-America: Mexico. *Human Biology* 68(3): 395–404

Madrigal, L. 1989. Hemoglobin genotype, fertility, and the malaria hypothesis. *Human Biology* 61: 311–25

Madrigal, L. 1991. The reliability of recalled estimates of menarcheal age in a sample of older women. *American Journal of Human Biology* 3: 105–10

Madrigal, L. 1994. Mortality seasonality in Escazu, Costa Rica, 1851–1921. *Human Biology,* 66(3): 433–52

Mascie-Taylor, C. G. N. and G. W. Lasker. 1993. *Research strategies in human biology: field and survey studies.* Cambridge studies in biological anthropology, N. 13: Cambridge University Press

Maxwell S. E. and H. D. Delayney. 1990. *Designing experiments and anlyzing data: A model comparison perspective.* Wadsworth publishing company: Belmont

Miller, G. R. and R. L. Burger. 1995. Our father the cayman, our dinner the llama: animal utilization at Chavín de Huántar, Peru. *American Antiquity* 60(3): 421–58

Mitchell R. J., L. Earl, P. Bray, Y. J. Fripp and J. Williams. 1994. DNA polymorphims at the lipoprotein lipase gene and their association with quantitative variation in plasma high-density lipoproteins and triacylglycerides. *Human Biology* 66(3): 383–97

Molnar S., C. Hildebolt, I. M. Molnar, J. Radovcic and M. Gravier. 1993. Hominid enamel thickness: I. The Krapina Neandertals. *American Journal of Physical Anthropology.* 92: 131–8

Neiman, F. D. 1995. Stylistic variation in evolutionary perspective: inferences from decorative diversity and interassemblage distance in Illinois Woodland ceramic assemblages. *American Antiquity* 60(1): 7–36

Oths, K. S. 1994. Health care decisions of households in economic crisis: an example from the Peruvian highlands. *Human Organization* 53(3): 245–53

Pollitzer W. S., M. T. Smith and W. R. Williams. 1988. A study of isonymic relationships in Fylingdales parish from marriage records from 1654 through 1916. *Human Biology* 60(3): 363–82

Purcell, T. W. 1993. *Banana fallout: Class, color and culture among West Indians in Costa Rica.* Center for Afro-American studies publications. University of California: Los Angeles

Quarles A., P. D. Williams, D. A. Hoyle, M. Brimeyer and A. R. Williams. 1994. Mothers' intention, age, education and the duration and management of breastfeeding. *Maternal-Child Nursing Journal* 22(3): 102–8

Reitz E. J., I. R. Quitmyer, H. S. Hale, S. J. Scudder, and E. S. Wing. 1987. Application of allometry to zooarchaeology. *American Antiquity,* 52(2): 304–17

Rosenbaum, P. R. 1995. *Observational studies*. Springer-Verlag: New York

SAS Institute Inc. 1991. *Getting started with the SAS System using SAS/AASIST software, version 6*, first edition, Cary: NC

SAS Institute Inc. 1992. *Doing more with SAS/ASSIST software, version 6*, first edition, Cary: NC

Shackley, M. S. 1995. Sources of archaeological obsidian in the greater American Southwest: an update and quantitative analysis. *American Antiquity* 60(3): 531–51

Smith J. J., L. Furbee, K. Maynard, S. Quick, and L. Ross. 1995. Salience counts: a domain analysis of English color terms. *Journal of Linguistic Anthropology* 5(2): 203–16

Sokal, R. R. and F. J. Rohlf. 1981. *Biometry*, second edition. W. H. Freeman and company: New York

Van Gerven D. P., S. G. Sheridan and W. Y. Adams. 1995. The health and nutrition of a medieval Nubian population. *American Anthropologist* 97 (3): 468–80

Volpe, E. P. 1985. *Understanding evolution*, fifth edition. Wm. C. Brown Publishers: Dubuque

White, N. M. 1994. Archaeological investigations at six sites in the Apalachicola river valley, Northwest Florida. A National oceanic and atmospheric administration technical memorandum (NOS SRD 26). US Department of Commerce, National Ocean Service: Washington, DC

White, T. D. 1991. *Human osteology*. Academic Press, Inc: San Diego

Wienker C. W. and K. A. Bennett. 1992. Trends and developments in physical anthropology, 1990–91. *American Journal of Physcial Anthropology* 87: 383–93

Chapter 1

3. $\Sigma X = 509$, $\Sigma X^2 = 23,197$, $(\Sigma X)^2 = 259,081$, $\Sigma Y = 1,256$, $\Sigma Y^2 = 116,970$, $(\Sigma Y)^2 = 1,577,536$, $\Sigma XY = 46,614$, $(\Sigma XY)^2 = 2,172,864,996$

Chapter 2

2.

Height	Frequency	%	Cumulative frequency	Cumulative %
167	1	5.3	1	5.3
176	1	5.3	2	10.5
179	1	5.3	3	15.8
180	1	5.3	4	21.1
183	1	5.3	5	26.3
185	1	5.3	6	31.6
187	2	10.5	8	42.1
188	1	5.3	9	47.4
189	3	15.8	12	63.2
190	2	10.5	14	73.7
192	1	5.3	15	78.9
193	1	5.3	16	84.2
194	1	5.3	17	89.5
197	1	5.3	18	94.7
199	1	5.3	19	100.0

Chapter 3

2. With the un-grouped data:

Range	Median	Sum	Mean	Variance	Std Dev.	Mode
32.0000	189	3554.00	187.0526316	57.7192982	7.5973218	189

Chapter 4

2. If $A = p = 0.92$ and $S = q = 0.08$, then the genotypic frequencies are: HbAA $= 0.8464$, HbAS $= 0.1472$ and HbSS $= 0.0064$.

4. (a) An individual with one or fewer lesions: $z = -5$, $p \approx 0$.

   (b) An individual with between three and six lesions: $z_1 = -2.5$, $z_2 = 1.25$, $p = 0.4938 + 0.3944 = 0.8882$.

   (c) An individual with between two and four lesions: $z_1 = 3.75$, $z_2 = 1.25$, $p = 0.499915 + 0.3944 = 0.1055$.

   (d). An individual with nine or more lesions: $z = 5$, $p \approx 0$.

Chapter 5

3. For a population with $\mu = 0$ and $\sigma = 1$, the $\alpha = 0.05$ cut-off point is 1.96.

   (a) $\sigma_{\bar{Y}} = 0.45$, $z = 0.89 \approx 0.9$. The area beyond this z score is $\beta = 0.4641$, so the power of the test is $1 - 0.4641 = 0.5359$.

   (b) $\sigma_{\bar{Y}} = 0.22$, $z = 0.18$. Thus: $\beta = 0.4286$, $1 - \beta = 0.5714$.

   (c) $\sigma_{\bar{Y}} = 0.45$, $z = 2.31$. Thus: $\beta = 0.0104$, $1 - \beta = 0.9896$.

   (d) $\sigma_{\bar{Y}} = 0.22$, $z = 4.73$. Thus: $\beta \approx 0$, $1 - \beta \approx 1$.

4. (a) For the z score: H0: $\mu = 18$, $\sigma = 2$, $n = 90$, $\bar{Y} = 20$. $\sigma_{\bar{Y}} = 0.21$, $z = 9.52$, $p < 0.05$. Reject H0.

   For the t score: H0: $\mu = 18$, $s = 2.3$, $n = 90$, $\bar{Y} = 20$. $s_{\bar{Y}} = 0.24$, $t = 8.3$, $p < 0.01$. Reject H0.

   (b) For the z score: H0: $\mu = 178$, $\sigma = 15$, $n = 300$, $\bar{Y} = 172$. $\sigma_{\bar{Y}} = 0.86$, $z = 6.97$, $p < 0.05$. Reject H0.

   For the t score: H0: $\mu = 178$, $s = 20$, $n = 300$, $\bar{Y} = 172$. $s_{\bar{Y}} = 1.15$, $t = 5.52$, $p < 0.01$. Reject H0.

Chapter 6

1. The assumptions of normality and homogeneity of variances are violated, so a non-parametric test is recommended. If the data had been normal, we would have used the t score for unequal variances, which indicates that the null hypothesis should not be rejected.

```
------------------------- LEVEL = 1 -------------------------
        W:Normal   0.826641   Pr<W    0.0001    Data not
------------------------- LEVEL = 2 -------------------------
        W:Normal   0.856181   Pr<W    0.0015    normal

TTEST PROCEDURE

Variances    T    DF    Prob>|T|
_____
Unequal  1.6942  45.9    0.0970
Equal    1.6284  54.0    0.1093
For H0: Variances are equal, F'=3.39 DF=(29,25) Prob>F'=0.0028
Variances not homogeneous_____↑
```

2. The differences are barely normal. If this is overlooked, we can reject the null hypothesis:

```
Variable=DIFF  W:Normal  0.914857  Pr<W  0.0653←almost significant

Mean            Std Error          T         Prob>|T|
------------------------------------------------------------
-1.0476190      0.3483033      -3.0077781     0.0070←Reject H0
------------------------------------------------------------
```

Chapter 7

1. The ANOVA table is:

```
General Linear Models Procedure
Dependent Variable: WT  Weight in grams
Source             DF    Sum of Squares   F Value    Pr>F
Model               2      147.95796296     2.94    0.0570
Error             105     2638.28500000
Corrected Total   107     2786.24296296
                General Linear Models Procedure

                T tests (LSD) for variable: WT
```

NOTE: This test controls the type I comparisonwise error rate
      not the experimentwise error rate.

```
            Alpha=0.05  df=105  MSE=25.12652
               Critical Value of T=1.98
           Least Significant Difference=2.3427
```

Means with the same letter are not significantly different.

```
            T Grouping    Mean   N  L
                    A    10.053   36  4
                    A
            B       A     9.064   36  3
            B
            B             7.228   36  1
            General Linear Models Procedure

            Scheffe's test for variable: WT
```

NOTE: This test controls the type I experimentwise error rate
      but generally has a higher type II error rate than
      REGWF for all pairwise comparisons

```
            Alpha=0.05 df=105 MSE=25.12652
               Critical Value of F=3.08285
           Minimum Significant Difference=2.9337
```

Means with the same letter are not significantly different.

```
            Scheffe Grouping    Mean   N  L
                    A         10.053   36  4
                    A
                    A          9.064   36  3
                    A
                    A          7.228   36  1
```

The ANOVA rejects the null hypothesis of equality of means. SAS was asked to perform the LSD test in addition to the Scheffe test, because the latter is so conservative that it did not reject the null hypothesis. According to the LSD test, the stratigraphic level #1 is significantly different from 4 and 3.

The assumptions of normality and homogeneity of variances were tested. The three groups suffered from non-normality:

```
------------------- Stratigraphic level=1 -------------------
     W:Normal   0.797614   Pr<W      0.0001← Data not normal
------------------- Stratigraphic level=3-------------------
     W:Normal   0.937973   Pr<W      0.0569← Almost significant
------------------- Stratigraphic level=4-------------------
     W:Normal   0.907947   Pr<W      0.0059← Data not normal
```

For the Fmax test, the two variances used were:

```
------------------- Stratigraphic level=1 -------------------
Std Dev   4.119751   Variance   16.97235
                              and
------------------- Stratigraphic level=4-------------------
Std Dev   6.274005   Variance   39.36313
```

Thus, Fmax: $39.36313/16.97235 = 2.32$. The critical value for $a = 3$ groups, and $n - 1 = 35$ ($n - 1 = 30$ is used instead) is 2.40. We do not reject the null hypothesis of homogeneity of variances.

Because of the non-normality of the samples, a non-parametric test should be used.

Chapter 8

1. The Mann–Whitney U test was done with SAS. The output is below:

```
          N P A R 1 W A Y   P R O C E D U R E

     Wilcoxon Scores (Rank Sums) for Variable AGE
            Classified by Variable GROUP

              Sum of   Expected     Std Dev      Mean
GROUP   N     Scores   Under H0     Under H0     Score
1       10     76.0      100.0    12.1611980   7.6000000
2        9    114.0       90.0    12.1611980  12.6666667
          Average Scores Were Used for Ties

Kruskal-Wallis Test (Chi-Square Approximation)
     CHISQ=3.8947 DF=1 Prob>CHISQ=0.0484← p value for H0
```

3. The archaeological problem was analyzed with a Kruskal-Wallis test. Below is the output.

```
            N P A R 1 W A Y   P R O C E D U R E

      Wilcoxon Scores (Rank Sums) for Variable WT
                 Classified by Variable L

            Sum of   Expected     Std Dev       Mean
   L   N    Scores   Under H0     Under H0      Score
   1  36 1641.00000    1962.0  153.412766  45.5833333
   3  36 2107.50000    1962.0  153.412766  58.5416667
   4  36 2137.50000    1962.0  153.412766  59.3750000
            Average Scores Were Used for Ties

   Kruskal-Wallis Test (Chi-Square Approximation)

   CHISQ=4.3909 DF=2 Prob>CHISQ=0.1113 ←p value for H0
```

4. There were four '0' differences, which were equally split between the negative and positive sums of ranks. The sum of the negative ranks was 22 and the sum of the positive ranks was 33. The critical value for $n = 10$ at $\alpha = 0.05$ is 8. The null hypothesis of no change is accepted.

Chapter 9

1. The linguistic data were analyzed using SAS. The entire analysis is not reproduced for space reasons. However, the steps outlined in the text are followed.

1. A plot of the data did not reveal a curved or fan-like behavior.
2. The equation was computed: $\hat{Y} = 5.16 + 0.14(X) - \epsilon$
3. The regression is expressed as an analysis of variance. The H0 that number of years does not explain variation of cartoon understanding is rejected. The Adj R-sq indicates that 0.3494 of the variation of cartoon understanding is explained by years of residence.

```
Model: MODEL1
Dependent Variable: CARTOONS

                     Analysis of Variance
                        Sum of       Mean
   Source      DF      Squares      Square  F Value
   Prob>F
   Model        1     55.65712    55.65712   14.960   0.0007
   Error       25     93.00955     3.72038           ↑p value
   C Total     26    148.66667

        Root MSE    1.92883   R-square    0.3744
        Dep Mean    7.22222   Adj R-sq    0.3494
        C.V.       26.70687
```

```
                     Parameter Estimates

                  Parameter   Standard    T for H0:
      Variable   DF  Estimate    Error      Parameter=0
      INTERCEP   1   5.156781   0.65034961  7.929
      YEARS      1   0.141182   0.03650169  3.868

      Variable   DF  Prob>|T|
      INTERCEP   1   0.0001
      YEARS      1   0.0007  ↔H0: β=0
```

4. The predicted values (and their confidence intervals) are computed:

Obs	Dep Var CARTOONS	Predict Value	Std Err Predict	Lower95% Mean	Upper95% Mean	Lower95% Predict
1	5.0000	6.1451	0.464	5.1893	7.1008	2.0592
2	7.0000	6.5686	0.408	5.7286	7.4086	2.5083
3	8.0000	7.2745	0.371	6.5095	8.0395	3.2290
4	10.0000	9.6746	0.735	8.1614	11.1878	5.4237
5	6.0000	7.1333	0.372	6.3674	7.8993	3.0877
6	8.0000	8.6863	0.530	7.5944	9.7782	4.5665
7	6.0000	6.4274	0.424	5.5536	7.3013	2.3599
8	5.0000	5.4391	0.592	4.2202	6.6581	1.2838
9	8.0000	7.4157	0.375	6.6443	8.1871	3.3690
10	10.0000	7.9804	0.420	7.1159	8.8450	3.9149
11	9.0000	6.1451	0.464	5.1893	7.1008	2.0592
12	8.0000	7.5569	0.381	6.7719	8.3419	3.5076
13	7.0000	6.7098	0.394	5.8980	7.5215	2.6552
14	10.0000	5.8627	0.511	4.8098	6.9156	1.7530
15	9.0000	10.8041	0.998	8.7493	12.8588	6.3316
16	10.0000	8.9687	0.585	7.7648	10.1726	4.8178
17	2.0000	6.1451	0.464	5.1893	7.1008	2.0592
18	4.0000	7.2745	0.371	6.5095	8.0395	3.2290
19	5.0000	5.4391	0.592	4.2202	6.6581	1.2838
20	2.0000	5.2980	0.621	4.0196	6.5764	1.1248
21	7.0000	9.3922	0.673	8.0067	10.7778	5.1851
22	6.0000	5.5803	0.564	4.4189	6.7417	1.4415
23	10.0000	8.8275	0.557	7.6807	9.9743	4.6928
24	7.0000	6.2862	0.443	5.3736	7.1989	2.2103
25	10.0000	7.8392	0.404	7.0071	8.6714	3.7805
26	6.0000	5.8627	0.511	4.8098	6.9156	1.7530
27	10.0000	8.2628	0.458	7.3186	9.2070	4.1796

5. The residuals are analyzed. None is particularly important.

Obs	Upper95% Predict	Residual	Std Err Residual	Student Residual	−2−1−0 1 2
1	10.2309	−1.1451	1.872	−0.612	| *| |
2	10.6289	0.4314	1.885	0.229	| | |

Obs	Upper95% Predict	Residual	Std Err Residual	Student Residual	−2−1−0 1 2
3	11.3200	0.7255	1.893	0.383	| | |
4	13.9255	0.3254	1.783	0.182	| | |
5	11.1790	−1.1333	1.893	−0.599	| * | |
6	12.8062	−0.6863	1.855	−0.370	| | |
7	10.4949	−0.4274	1.882	−0.227	| | |
8	9.5945	−0.4391	1.836	−0.239	| | |
9	11.4624	0.5843	1.892	0.309	| | |
10	12.0459	2.0196	1.883	1.073	| | ** |
11	10.2309	2.8549	1.872	1.525	| | *** |
12	11.6062	0.4431	1.891	0.234	| | |
13	10.7644	0.2902	1.888	0.154	| | |
14	9.9723	4.1373	1.860	2.225	| | ****|
15	15.2765	−1.8041	1.651	−1.093	| ** | |
16	13.1196	1.0313	1.838	0.561	| | * |
17	10.2309	−4.1451	1.872	−2.214	|****| |
18	11.3200	−3.2745	1.893	−1.730	| *** | |
19	9.5945	−0.4391	1.836	−0.239	| | |
20	9.4711	−3.2980	1.826	−1.806	| *** | |
21	13.5994	−2.3922	1.808	−1.323	| ** | |
22	9.7191	0.4197	1.845	0.228	| | |
23	12.9622	1.1725	1.847	0.635	| | * |
24	10.3622	0.7138	1.877	0.380	| | |
25	11.8980	2.1608	1.886	1.146	| | ** |
26	9.9723	0.1373	1.860	0.074	| | |
27	12.3459	1.7372	1.874	0.927	| | * |

A normality test of the residuals indicates that they are normal:

```
Univariate Procedure

Variable=R          Residual
                  Moments

N                 27    Sum Wgts          27
Mean               0    Sum                0
Std Dev    1.891373    Variance     3.57729
Skewness  −0.26813     Kurtosis    0.355029
USS        93.00955    CSS         93.00955
CV                .    Std Mean    0.363995
T:Mean=0           0    Pr>|T|       1.0000
Num ^=0           27    Num>0             16
M(Sign)          2.5    Pr>=|M|      0.4421
Sgn Rank          10    Pr>=|S|      0.8153

W:Normal    0.971534   Pr<W        0.6622←Normal
```

A histogram of the residuals also indicates that they are normally distributed.

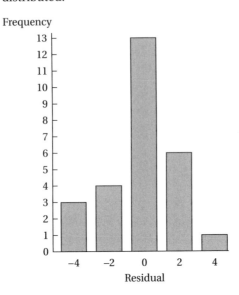

Chapter 10

1. The abuse data were entered into SAS. First, the data plot was inspected to determine if the data violated the assumptions of correlation analysis. The plot indicated that the data suffer from heterocedasticity (there was more variation on the low income range).

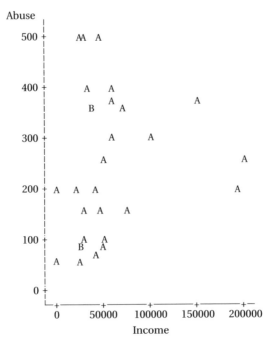

A Spearman correlation is probably more appropriate, or a transformation should be attempted. However, the Pearson correlation analysis is:

```
                        Correlation Analysis

                2 'VAR' Variables: ABUSE INCOME
      Pearson Correlation Coefficients/Prob>|R| under H0: Rho=0/N=29
                    ABUSE      INCOME
          ABUSE    1.00000   0.13699
                   0.0       0.4786←p value. Accept H0
          INCOME   0.13699   1.00000
                   0.4786    0.0
```

Because of the concerns about heterocedasticity, the Spearman correlation was also applied to the data. The output is below:

```
                        Correlation Analysis

      Spearman Correlation Coefficients/Prob >|R| under H0: Rho=0
      /N=29
                    ABUSE      INCOME
          ABUSE    1.00000   0.28600
                   0.0       0.1326←p value. Accept H0
          INCOME   0.28600   1.00000
                   0.1326    0.0
```

Chapter 11

2. The gender by status problem was analyzed using SAS. Below is the X^2 statistic:

```
STATISTICS FOR TABLE OF STATUS BY GENDER
Statistic          DF   Value    Prob
--------------------------------------------------
Chi-Square          1   2.705    0.100←Accept H0
Sample Size=64
```

The purpose of this appendix is to give readers who choose to use SAS/ASSIST in conjunction with this book an introduction to the package. However, this appendix does not intend to be, nor can it be, a substitute for a manual (SAS Institute Inc. 1991, 1992). Readers should also be aware that the specific set-up in their machine may differ from the one illustrated in this book. For example, the tasks performed by function keys may differ, or the tasks available in menus may differ. Another aspect not covered here is how to create permanent SAS libraries, a place in the computer's memory where data or programs are stored permanently. It is assumed here that the user knows how to start SAS, *and* how to get to SAS/ASSIST.

The reason SAS/ASSIST was used to illustrate computer statistical calculations in this book is that it is an extremely user-friendly, menu-driven application. The user does not have to write any programs, but simply enters a data set, and requests the statistical analysis from menus. When used in the textbook, SAS/ASSIST was illustrated by explaining how the data were entered, and by listing the menu tasks which were chosen. Because SAS/ASSIST is so easy to learn to use, the fact that the readers' set-up may differ slightly from the one illustrated in the book is not a major problem. The user is encouraged to navigate throughout, and experience the various menus for a while. This, in conjunction with the tutorial provided by SAS/ASSIST, is probably the best way to gain familiarity with the program.

The negative aspect of such an easy-to-use application is that some complex tasks are not available through the menus. In that case, the user needs to write the appropriate code in the program editor. When it was necessary to write a SAS program to perform a statistical test in the book, the code was reproduced for the readers' benefit. However, the rules of SAS syntax were not reviewed at all, since that would fall outside the primary purpose of the book. After all, many readers may choose to use a different computer package, or none at all.

One of the most attractive aspects of SAS/ASSIST is that it has available under the help menu a list of sample programs which can be copied and pasted into the program editor, from where they are executed. Users can then recall the program, and adapt it to their own needs, redirecting SAS to their data set, changing titles, etc. Therefore, if a task is not available through the menus, the user can use one of these programs, adapt it, run it and save it.

If the user desires to perform a task which is not obviously accessible by any of the menus, he can also click on the index and look for the task in an alphabetized list. If the desired statistical test is in the list then, by clicking on it, the user will be taken to a menu which offers the task. In this case, the user does not have to write the program either.

This appendix illustrates in a brief manner the use of SAS/ASSIST in three ways: menu-driven tasks, transforming a sample program, and using the index.

Menu-driven tasks. We will practice the use of menus by entering a small data set (2, 4, 9, 12), and by computing a few of its descriptive statistics. The first step is to enter the data, which we do by clicking on: 1. data management, 2. create/import, 3. enter data interactively, 4. enter data in tabular form. The next window asks the user to name the data set. We name the data 'temp', and put it in the temporary library. SAS then asks the user to name and give other information about the variables to be entered. If the data are characters, then a '$' sign must be entered under the 'type' heading. If numbers are entered, the only thing that must be typed is the name of the variable, which we will call 'numbers'. When done, the user pulls down the File menu, and clicks on 'end'. Now the user enters the data and, when done, pulls down the File menu and clicks on 'end'. By clicking on the 'go back' button, the user gets back to the primary menu. To obtain descriptive statistics of the sample, the following tasks are chosen: 1. data analysis, 2. elementary, 3. summary statistics. At this window, choose the mean and the standard deviation. In general, to run a menu-driven task, the user pulls down the locals menu, and clicks on 'run'. The output appears in the window.

```
Analysis Variable : NUMBERS

      Mean      Std Dev
-------------------------
6.7500000  4.5734742
-------------------------
```

One of the most attractive aspects of SAS/ASSIST is that it allows the user to retrieve the program used in the menu-driven task, and change it. For example, the user may want to change the default width and length of the output page (something I did repeatedly for this book, since I would copy the program and paste it in the word-processor file). To do this, the user would go to the Windows menu, click on program manager, and recall the program (function 4 in my set-up). After making the desired changes, the user may run the program (by clicking on the running figure icon), and even save it. The user would now be working with the more traditional SAS PC as opposed to SAS/ASSIST.

Transforming a sample program. The list of companion programs can be accessed from any window, under the Help menu. The user pulls down the Help menu, and clicks on 'sample programs'. A list of the major SAS tasks (e.g. graphics, ETS, statistics, etc.) appears. If the user does not know where the task she wishes to perform is, she can search the list by clicking on 'search', or she can browse through and choose from a list of available programs. A program can be chosen, opened, and copied by pulling down the Edit menu. Then, the windows can be changed to the program-editor window, where the program is pasted. The program is run by clicking on the running figure icon, and the output appears in the window. If the program does what the user wants, she can recall the program and change it so that it works with her data, produces the appropriate titles, etc. If the user wishes, the program can be saved for later use.

Using the index. If the user does not find the desired task in the menus, then she may click on the index, and attempt to find it in the list. After clicking on the task, the window changes to a window in which SAS is directed to a data set. At this point, the user is working with the menus, and would follow the usual steps.

For readers who have used the more traditional versions of SAS, or any other package in which a program was written, the approach of SAS/ASSIST will be quite different. These users will have to learn a very different approach to computer statistical analysis. However, they will benefit from their previous experience by being able to write or alter programs if the need arises. Individuals who have never used a statistical computer package will be pleased by the ease with which they obtain their results. However, if they desire to compute more sophisticated statistical tests, they may need to read about the SAS system's rules on programming.

Table 1 The unit normal table

Column A lists the z score values.
Column B provides the proportion of area between the mean and the z score value.
Column C provides the proportion of area beyond the z score.

(A)	(B) Area between mean and z	(C) Area beyond z	(A)	(B) Area between mean and z	(C) Area beyond z	(A)	(B) Area between mean and z	(C) Area beyond z
0.00	.0000	.5000	0.29	.1141	.3859	0.58	.2190	.2810
0.01	.0040	.4960	0.30	.1179	.3821	0.59	.2224	.2776
0.02	.0080	.4920	0.31	.1217	.3783	0.60	.2257	.2743
0.03	.0120	.4880	0.32	.1255	.3745	0.61	.2291	.2709
0.04	.0160	.4840	0.33	.1293	.3707	0.62	.2324	.2676
0.05	.0199	.4801	0.34	.1331	.3669	0.63	.2357	.2643
0.06	.0239	.4761	0.35	.1368	.3632	0.64	.2389	.2611
0.07	.0279	.4721	0.36	.1406	.3594	0.65	.2422	.2578
0.08	.0319	.4681	0.37	.1443	.3557	0.66	.2454	.2546
0.09	.0359	.4641	0.38	.1480	.3520	0.67	.2486	.2514
0.10	.0398	.4602	0.39	.1517	.3483	0.68	.2517	.2483
0.11	.0438	.4562	0.40	.1554	.3446	0.69	.2549	.2451
0.12	.0478	.4522	0.41	.1591	.3409	0 70	.2580	.2420
0.13	.0517	.4483	0.42	.1628	.3372	0.71	.2611	.2389
0.14	.0557	.4443	0.43	.1664	.3336	0.72	.2642	.2358
0.15	.0596	.4404	0.44	.1700	.3300	0.73	.2673	.2327
0.16	.0636	.4364	0.45	.1736	.3264	0.74	.2704	.2296
0.17	.0675	.4325	0.46	.1772	.3228	0.75	.2734	.2266
0.18	.0714	.4286	0.47	.1808	.3192	0.76	.2674	.2236
0.19	.0753	.4247	0.48	.1844	.3156	0.77	.2794	.2206
0.20	.0793	.4207	0.49	.1879	.3121	0.78	.2823	.2177
0.21	.0832	.4168	0.50	.1915	.3085	0.79	.2852	.2148
0.22	.0871	.4129	0.51	.1950	.3050	0.80	.2881	.2119
0.23	.0910	.4090	0.52	.1985	.3015	0.81	.2910	.2090
0.24	.0948	.4052	0.53	.2019	.2981	0.82	.2939	.2061
0.25	.0987	.4013	0.54	.2054	.2946	0.83	.2967	.2033
0.26	.1026	.3974	0.55	.2088	.2912	0.84	.2995	.2005
0.27	.1064	.3936	0.56	.2123	.2877	0.85	.3023	.1977
0.28	.1103	.3897	0.57	.2157	.2843	0.86	.3051	.1949

Table 1 (*cont.*)

(A) z	(B) Area between mean and z	(C) Area beyond z	(A)	(B) Area between mean and z	(C) Area beyond z	(A)	(B) Area between mean and z	(C) Area beyond z
0.87	.3078	.1922	1.31	.4049	.0951	1.75	.4599	.0401
0.88	.3106	.1894	1.32	.4066	.0934	1.76	.4608	.0392
0.89	.3133	.1867	1.33	.4082	.0918	1.77	.4616	.0384
0.90	.3159	.1841	1.34	.4099	.0901	1.78	.4625	.0375
0.91	.3186	.1814	1.35	.4115	.0885	1.79	.4633	.0367
0.92	.3212	.1788	1.36	.4131	.0869	1.80	.4641	.0359
0.93	.3238	.1762	1.37	.4147	.0853	1.81	.4649	.0351
0.94	.3264	.1736	1.38	.4162	.0838	1.82	.4656	.0344
0.95	.3289	.1711	1.39	.4177	.0823	1.83	.4664	.0336
0.96	.3315	.1685	1.40	.4192	.0808	1.84	.4671	.0329
0.97	.3340	.1660	1.41	.4207	.0793	1.85	.4678	.0322
0.98	.3365	.1635	1.42	.4222	.0778	1.86	.4686	.0314
0.99	.3389	.1611	1.43	.4236	.0764	1.87	.4693	.0307
1.00	.3413	.1587	1.44	.4251	.0749	1.88	.4699	.0301
1.01	.3438	.1562	1.45	.4265	.0735	1.89	.4706	.0294
1.02	.3461	.1539	1.46	.4279	.0721	1.90	.4713	.0287
1.03	.3485	.1515	1.47	.4292	.0708	1.91	.4719	.0281
1.04	.3508	.1492	1.48	.4306	.0694	1.92	.4726	.0274
1.05	.3531	.1469	1.49	.4319	.0681	1.93	.4732	.0268
1.06	.3554	.1446	1.50	.4332	.0668	1.94	.4738	.0262
1.07	.3577	.1423	1.51	.4345	.0655	1.95	.4744	.0256
1.08	.3599	.1401	1.52	.4357	.0643	1.96	.4750	.0250
1.09	.3621	.1379	1.53	.4370	.0630	1.97	.4756	.0244
1.10	.3643	.1357	1.54	.4382	.0618	1.98	.4761	.0239
1.11	.3665	.1335	1.55	.4394	.0606	1.99	.4767	.0233
1.12	.3686	.1314	1.56	.4406	.0594	2.00	.4772	.0228
1.13	.3708	.1292	1.57	.4418	.0582	2.01	.4778	.0222
1.14	.3729	.1271	1.58	.4429	.0571	2.02	.4783	.0217
1.15	.3749	.1251	1.59	.4441	.0559	2.03	.4788	.0212
1.16	.3770	.1230	1.60	.4452	.0548	2.04	.4793	.0207
1.17	.3790	.1210	1.61	.4463	.0537	2.05	.4798	.0202
1.18	.3810	.1190	1.62	.4474	.0526	2.06	.4803	.0197
1.19	.3830	.1170	1.63	.4484	.0516	2.07	.4808	.0192
1.20	.3849	.1151	1.64	.4495	.0505	2.08	.4812	.0188
1.21	.3859	.1131	1.65	.4505	.0495	2.09	.4817	.0183
1.22	.3888	.1112	1.66	.4515	.0485	2.10	.4821	.0179
1.23	.3907	.1093	1.67	.4525	.0475	2.11	.4826	.0174
1.24	.3925	.1075	1.68	.4535	.0465	2.12	.4830	.0170
1.25	.3944	.1056	1.69	.4545	.0455	2.13	.4834	.0166
1.26	.3962	.1038	1.70	.4554	.0446	2.14	.4838	.0162
1.27	.3980	.1020	1.71	.4564	.0436	2.15	.4842	.0158
1.28	.3997	.1003	1.72	.4573	.0427	2.16	.4846	.0154
1.29	.4015	.0985	1.73	.4582	.0418	2.17	.4850	.0150
1.30	.4032	.0968	1.74	.4591	.0409	2.18	.4854	.0146

Table 1 (*cont.*)

(A) z	(B) Area between mean and z	(C) Area beyond z	(A) z	(B) Area between mean and z	(C) Area beyond z	(A) z	(B) Area between mean and z	(C) Area beyond z
2.19	.4857	.0143	2.57	.4949	.0051	2.95	.4984	.0016
2.20	.4861	.0139	2.58	.4951	.0049	2.96	.4985	.0015
2.21	.4864	.0136	2.59	.4952	.0048	2.97	.4985	.0015
2.22	.4868	.0132	2.60	.4953	.0047	2.98	.4986	.0014
2.23	.4871	.0129	2.61	.4955	.0045	2.99	.4986	.0014
2.24	.4875	.0125	2.62	.4956	.0044	3.00	.4987	.0013
2.25	.4878	.0122	2.63	.4957	.0043	3.01	.4987	.0013
2.26	.4881	.0119	2.64	.4959	.0041	3.02	.4987	.0013
2.27	.4884	.0116	2.65	.4960	.0040	3.03	.4988	.0012
2.28	.4887	.0113	2.66	.4961	.0039	3.04	.4988	.0012
2.29	.4890	.0110	2.67	.4962	.0038	3.05	.4989	.0011
2.30	.4893	.0107	2.68	.4963	.0037	3.06	.4989	.0011
2.31	.4896	.0104	2.69	.4964	.0036	3.07	.4989	.0011
2.32	.4898	.0102	2.70	.4965	.0035	3.08	.4990	.0010
2.33	.4901	.0099	2.71	.4966	.0034	3.09	.4990	.0010
2.34	.4904	.0096	2.72	.4967	.0033	3.10	.4900	.0010
2.35	.4906	.0094	2.73	.4968	.0032	3.11	.4991	.0009
2.36	.4909	.0091	2.74	.4969	.0031	3.12	.4991	.0009
2.37	.4911	.0089	2.75	.4970	.0030	3.13	.4991	.0009
2.38	.4913	.0087	2.76	.4971	.0029	3.14	.4992	.0008
2.39	.4916	.0084	2.77	.4972	.0028	3.15	.4992	.0008
2.40	.4918	.0082	2.78	.4973	.0027	3.16	.4992	.0008
2.41	.4920	.0080	2.79	.4974	.0026	3.17	.4992	.0008
2.42	.4922	.0078	2.80	.4974	.0026	3.18	.4993	.0007
2.43	.4925	.0075	2.81	.4975	.0025	3.19	.4993	.0007
2.44	.4927	.0073	2.82	.4976	.0024	3.20	.4993	.0007
2.45	.4929	.0071	2.83	.4977	.0023	3.21	.4993	.0007
2.46	.4931	.0069	2.84	.4977	.0023	3.22	.4994	.0006
2.47	.4932	.0068	2.85	.4978	.0022	3.23	.4994	.0006
2.48	.4934	.0066	2.86	.4979	.0021	3.24	.4994	.0006
2.49	.4936	.0064	2.87	.4979	.0021	3.30	.4995	.0005
2.50	.4938	.0062	2.88	.4980	.0020	3.40	.4997	.0003
2.51	.4940	.0060	2.89	.4981	.0019	3.50	.4998	.0002
2.52	.4941	.0059	2.90	.4981	.0019	3.60	.4998	.0002
2.53	.4943	.0057	2.91	.4982	.0018	3.70	.4999	.0001
2.54	.4945	.0055	2.92	.4982	.0018	3.80	.49993	.00007
2.55	.4946	.0054	2.93	.4983	.0017	3.90	.49995	.00005
2.56	.4948	.0052	2.94	.4984	.0016	4.00	.49997	.00003

Note:
Because the normal distribution is symmetrical, areas for negative z scores are the same as those for positive z scores.

Source: Copyright permission requested from Macmillan Publishing Company. The table appeared as Appendix 2 on pages 373–377 of *Statistical reasoning and procedures*, by R. B. Clarke, A. P. Coladarci and J. Caffrey, 1965, published by Charles E. Merril.

Table 2 Critical values of the *t* distribution

	Proportion in one tail					
	0.25	0.10	0.05	0.025	0.01	0.005
	Proportion in two tails					
df	0.50	0.20	0.10	0.05	0.02	0.01
1	1.000	3.078	6.314	12.706	31.821	63.657
2	0.816	1.886	2.920	4.303	6.965	9.925
3	0.765	1.638	2.353	3.182	4.541	5.841
4	0.741	1.533	2.132	2.776	3.747	4.604
5	0.727	1.476	2.015	2.571	3.365	4.032
6	0.718	1.440	1.943	2.447	3.143	3.707
7	0.711	1.415	1.895	2.365	2.998	3.499
8	0.706	1.397	1.860	2.306	2.896	3.355
9	0.703	1.383	1.833	2.262	2.821	3.250
10	0.700	1.372	1.812	2.228	2.764	3.169
11	0.697	1.363	1.796	2.201	2.718	3.106
12	0.695	1.356	1.782	2.179	2.681	3.055
13	0.694	1.350	1.771	2.160	2.650	3.012
14	0.692	1.345	1.761	2.145	2.624	2.977
15	0.691	1.341	1.753	2.131	2.602	2.947
16	0.690	1.337	1.746	2.120	2.583	2.921
17	0.689	1.333	1.740	2.110	2.567	2.898
18	0.688	1.330	1.734	2.101	2.552	2.878
19	0.688	1.328	1.729	2.093	2.539	2.861
20	0.687	1.325	1.725	2.086	2.528	2.845
21	0.686	1.323	1.721	2.080	2.518	2.831
22	0.686	1.321	1.717	2.074	2.508	2.819
23	0.685	1.319	1.714	2.069	2.500	2.807
24	0.685	1.318	1.711	2.064	2.492	2.797
25	0.684	1.316	1.708	2.060	2.485	2.787
26	0.684	1.315	1.706	2.056	2.479	2.779
27	0.684	1.314	1.703	2.052	2.473	2.771
28	0.683	1.313	1.701	2.048	2.467	2.763
29	0.683	1.311	1.699	2.045	2.462	2.756
30	0.683	1.310	1.697	2.042	2.457	2.750
40	0.681	1.303	1.684	2.021	2.423	2.704
60	0.679	1.296	1.671	2.000	2.390	2.660
120	0.677	1.289	1.658	1.980	2.358	2.617
∞	0.674	1.282	1.645	1.960	2.326	2.576

Source: Reprinted by permission of Addison Wesley Longman Ltd. The table appeared as table III (Distribution of *t*) on page 46 of *Statistical tables for biological, agricultural and medical researcher*, sixth edition, by R. A. Fisher and F. Yates, 1974.

Table 3 Upper 5 and 1% points of the maximum *F*-ratio

$N-1$ \ a	2	3	4	5	6	7	8	9	10	11	12
2	39.0	87.5	142.	202.	266.	333.	403.	475.	550.	626.	704.
	199.	448.	729.	1036.	1362.	1705.	2063.	2432.	2813.	3204.	3605.
3	15.4	27.8	39.2	50.7	62.0	72.9	83.5	93.9	104.	114.	124.
	47.5	85.	120	151.	184.	21(6)	24(9)	28(1)	31(0)	33(7)	36(1)
4	9.60	15.5	20.6	25.2	29.5	33.6	37.5	41.1	44.6	48.0	51.4
	23.2	37.	49.	59.	69.	79.	89.	97.	106.	113.	120.
5	7.15	10.8	13.7	16.3	18.7	20.8	22.9	24.7	26.5	28.2	29.9
	14.9	22.	28.	33.	38.	42.	46.	50.	54.	57.	60.
6	5.82	8.38	10.4	12.1	13.7	15.0	16.3	17.5	18.6	19.7	20.7
	11.1	15.5	19.1	22.	25.	27.	30.	32.	34.	36.	37.
7	4.99	6.94	8.44	9.70	10.8	11.8	12.7	13.5	14.3	15.1	15.8
	8.89	12.1	14.5	16.5	18.4	20.	22.	23.	24.	26.	27.
8	4.43	6.00	7.18	8.12	9.03	9.78	10.5	11.1	11.7	12.2	12.7
	7.50	9.9	11.7	13.2	14.5	15.8	16.9	17.9	18.9	19.8	21.
9	4.03	5.34	6.31	7.11	7.80	8.41	8.95	9.45	9.91	10.3	10.7
	6.54	8.5	9.9	11.1	12.1	13.1	13.9	14.7	15.3	16.0	16.6
10	3.72	4.85	5.67	6.34	6.92	7.42	7.87	8.28	8.66	9.01	9.34
	5.85	7.4	8.6	9.6	10.4	11.1	11.8	12.4	12.9	13.4	13.9
12	3.28	4.16	4.79	5.30	5.72	6.09	6.42	6.72	7.00	7.25	7.48
	4.91	6.1	6.9	7.6	8.2	8.7	9.1	9.5	9.9	10.2	10.6
15	2.86	3.54	4.01	4.37	4.68	4.95	5.19	5.40	5.59	5.77	5.93
	4.07	4.9	5.5	6.0	6.4	6.7	7.1	7.3	7.5	7.8	8.0
20	2.46	2.95	3.29	3.54	3.76	3.49	4.10	4.24	4.37	4.49	4.59
	3.32	3.8	4.3	4.6	4.9	5.1	5.3	5.5	5.6	5.8	5.9
30	2.07	2.40	2.61	2.78	2.91	3.02	3.12	3.21	3.29	3.36	3.39
	2.63	3.0	3.3	3.4	3.6	3.7	3.8	3.9	4.0	4.1	4.2
60	1.67	1.85	1.96	2.04	2.11	2.17	2.22	2.26	2.30	2.33	2.36
	1.96	2.2	2.3	2.4	2.4	2.5	2.5	2.6	2.6	2.7	2.7
∞	1.00	1.00	1.00	1.00	1.00	1.00	1.00	1.00	1.00	1.00	1.00
	1.00	1.00	1.00	1.00	1.00	1.00	1.00	1.00	1.00	1.00	1.00

Source: Reprinted with permission from the Biometrika Trustees. Table appeared on p. 424 of the paper by H. A. David 'Upper 5 and 1% points of the maximum *F*-ratio', published in *Biometrika*, vol. 39, 1952.

Table 4 Critical values of the F distribution

Degrees freedom: denominator	Degrees of freedom: numerator														
	1	2	3	4	5	6	7	8	9	10	11	12	14	16	20
1	161	200	216	225	230	234	237	239	241	242	243	244	245	246	248
	4052	**4999**	**5403**	**5625**	**5764**	**5859**	**5928**	**5981**	**6022**	**6056**	**6082**	**6106**	**6142**	**6169**	**6208**
2	18.51	19.00	19.16	19.25	19.30	19.33	19.36	19.37	19.38	19.39	19.40	19.41	19.42	19.43	19.44
	98.49	**99.00**	**99.17**	**99.25**	**99.30**	**99.33**	**99.34**	**99.36**	**99.38**	**99.40**	**99.41**	**99.41**	**99.43**	**99.44**	**99.45**
3	10.13	9.55	9.28	9.12	9.01	8.94	8.88	8.84	8.81	8.78	8.76	8.74	8.71	8.69	8.66
	34.12	**30.82**	**29.46**	**28.71**	**28.24**	**27.91**	**27.67**	**27.49**	**27.34**	**27.23**	**27.13**	**27.05**	**26.92**	**26.83**	**26.69**
4	7.71	6.94	6.59	6.39	6.26	6.16	6.09	6.04	6.00	5.96	5.93	5.91	5.87	5.84	5.80
	21.20	**18.00**	**16.69**	**15.98**	**15.52**	**15.21**	**14.98**	**14.80**	**14.66**	**14.54**	**14.45**	**14.37**	**14.24**	**14.15**	**14.02**
5	6.61	5.79	5.41	5.19	5.05	4.95	4.88	4.82	4.78	4.74	4.70	4.68	4.64	4.60	4.56
	16.26	**13.27**	**12.06**	**11.39**	**10.97**	**10.67**	**10.45**	**10.27**	**10.15**	**10.05**	**9.96**	**9.89**	**9.77**	**9.68**	**9.55**
6	5.99	5.14	4.76	4.53	4.39	4.28	4.21	4.15	4.10	4.06	4.03	4.00	3.96	3.92	3.87
	13.74	**10.92**	**9.78**	**9.15**	**8.75**	**8.47**	**8.26**	**8.10**	**7.98**	**7.87**	**7.79**	**7.72**	**7.60**	**7.52**	**7.39**
7	5.59	4.47	4.35	4.12	3.97	3.87	3.79	3.73	3.68	3.63	3.60	3.57	3.52	3.49	3.44
	12.25	**9.55**	**8.45**	**7.85**	**7.46**	**7.19**	**7.00**	**6.84**	**6.71**	**6.62**	**6.54**	**6.47**	**6.35**	**6.27**	**6.15**
8	5.32	4.46	4.07	3.84	3.69	3.58	3.50	3.44	3.39	3.34	3.31	3.28	3.23	3.20	3.15
	11.26	**8.65**	**7.59**	**7.01**	**6.63**	**6.37**	**6.19**	**6.03**	**5.91**	**5.82**	**5.74**	**5.67**	**5.56**	**5.48**	**5.36**
9	5.12	4.26	3.86	3.63	3.48	3.37	3.29	3.23	3.18	3.13	3.10	3.07	3.02	2.98	2.93
	10.56	**8.02**	**6.99**	**6.42**	**6.06**	**5.80**	**5.62**	**5.47**	**5.35**	**5.26**	**5.18**	**5.11**	**5.00**	**4.92**	**4.80**
10	4.96	4.10	3.71	3.48	3.33	3.22	3.14	3.07	3.02	2.97	2.94	2.91	2.86	2.82	2.77
	10.04	**7.56**	**6.55**	**5.99**	**5.64**	**5.39**	**5.21**	**5.06**	**4.95**	**4.85**	**4.78**	**4.71**	**4.60**	**4.52**	**4.41**
11	4.84	3.98	3.59	3.36	3.20	3.09	3.01	2.95	2.90	2.86	2.82	2.79	2.74	2.70	2.65
	9.65	**7.20**	**6.22**	**5.67**	**5.32**	**5.07**	**4.88**	**4.74**	**4.63**	**4.54**	**4.46**	**4.40**	**4.29**	**4.21**	**4.10**
12	4.75	3.88	3.49	3.26	3.11	3.00	2.92	2.85	2.80	2.76	2.72	2.69	2.64	2.60	2.54
	9.33	**6.93**	**5.95**	**5.41**	**5.06**	**4.82**	**4.65**	**4.50**	**4.39**	**4.30**	**4.22**	**4.16**	**4.05**	**3.98**	**3.86**
13	4.67	3.80	3.41	3.18	3.02	2.92	2.84	2.77	2.72	2.67	2.63	2.60	2.55	2.51	2.46
	9.07	**6.70**	**5.74**	**5.20**	**4.86**	**4.62**	**4.44**	**4.30**	**4.19**	**4.10**	**4.02**	**3.96**	**3.85**	**3.78**	**3.67**
14	4.60	3.74	3.34	3.11	2.96	2.85	2.77	2.70	2.65	2.60	2.56	2.53	2.48	2.44	2.39

Note: Table entries in lightface type are critical values for the .05 level of significance. Boldface type values are for the .01 level of significance.

Table 4 (cont.)

Degrees freedom: denominator		Degrees of freedom: numerator														
		1	2	3	4	5	6	7	8	9	10	11	12	14	16	20
15		**8.86**	**6.51**	**5.56**	**5.03**	**4.69**	**4.46**	**4.28**	**4.14**	**4.03**	**3.94**	**3.86**	**3.80**	**3.70**	**3.62**	**3.51**
		4.54	3.68	3.29	3.06	2.90	2.79	2.70	2.64	2.59	2.55	2.51	2.48	2.43	2.39	2.33
16		**8.68**	**6.36**	**5.42**	**4.89**	**4.56**	**4.32**	**4.14**	**4.00**	**3.89**	**3.80**	**3.73**	**3.67**	**3.56**	**3.48**	**3.36**
		4.49	3.63	3.24	3.01	2.85	2.74	2.66	2.59	2.54	2.49	2.45	2.42	2.37	2.33	2.28
17		**8.53**	**6.23**	**5.29**	**4.77**	**4.44**	**4.20**	**4.03**	**3.89**	**3.78**	**3.69**	**3.61**	**3.55**	**3.45**	**3.37**	**3.25**
		4.45	3.59	3.20	2.96	2.81	2.70	2.62	2.55	2.50	2.45	2.41	2.38	2.33	2.29	2.23
18		**8.40**	**6.11**	**5.18**	**4.67**	**4.34**	**4.10**	**3.93**	**3.79**	**3.68**	**3.59**	**3.52**	**3.45**	**3.35**	**3.27**	**3.16**
		4.41	3.55	3.16	2.93	2.77	2.66	2.58	2.51	2.46	2.41	2.37	2.34	2.29	2.25	2.19
19		**8.28**	**6.01**	**5.09**	**4.58**	**4.25**	**4.01**	**3.85**	**3.71**	**3.60**	**3.51**	**3.44**	**3.37**	**3.27**	**3.19**	**3.07**
		4.38	3.52	3.13	2.90	2.74	2.63	2.55	2.48	2.43	2.38	2.34	2.31	2.26	2.21	2.15
20		**8.18**	**5.93**	**5.01**	**4.50**	**4.17**	**3.94**	**3.77**	**3.63**	**3.52**	**3.43**	**3.36**	**3.30**	**3.19**	**3.12**	**3.00**
		4.35	3.49	3.10	2.87	2.71	2.60	2.52	2.45	2.40	2.35	2.31	2.28	2.23	2.18	2.12
21		**8.10**	**5.85**	**4.94**	**4.43**	**4.10**	**3.87**	**3.71**	**3.56**	**3.45**	**3.37**	**3.30**	**3.23**	**3.13**	**3.05**	**2.94**
		4.32	3.47	3.07	2.84	2.68	2.57	2.49	2.42	2.37	2.32	2.28	2.25	2.20	2.15	2.09
22		**8.02**	**5.78**	**4.87**	**4.37**	**4.04**	**3.81**	**3.65**	**3.51**	**3.40**	**3.31**	**3.24**	**3.17**	**3.07**	**2.99**	**2.88**
		4.30	3.44	3.05	2.82	2.66	2.55	2.47	2.40	2.35	2.30	2.26	2.23	2.18	2.13	2.07
23		**7.94**	**5.72**	**4.82**	**4.31**	**3.99**	**3.76**	**3.59**	**3.45**	**3.35**	**3.26**	**3.18**	**3.12**	**3.02**	**2.94**	**2.83**
		4.28	3.42	3.03	2.80	2.64	2.53	2.45	2.38	2.32	2.28	2.24	2.20	2.14	2.10	2.04
24		**7.88**	**5.66**	**4.76**	**4.26**	**3.94**	**3.71**	**3.54**	**3.41**	**3.30**	**3.21**	**3.14**	**3.07**	**2.97**	**2.89**	**2.78**
		4.26	3.40	3.01	2.78	2.62	2.51	2.43	2.36	2.30	2.26	2.22	2.18	2.13	2.09	2.02
25		**7.82**	**5.61**	**4.72**	**4.22**	**3.90**	**3.67**	**3.50**	**3.36**	**3.25**	**3.17**	**3.09**	**3.03**	**2.93**	**2.85**	**2.74**
		4.24	3.38	2.99	2.76	2.60	2.49	2.41	2.34	2.28	2.24	2.20	2.16	2.11	2.06	2.00
26		**7.77**	**5.57**	**4.68**	**4.18**	**3.86**	**3.63**	**3.46**	**3.32**	**3.21**	**3.13**	**3.05**	**2.99**	**2.89**	**2.81**	**2.70**
		4.22	3.37	2.98	2.74	2.59	2.47	2.39	2.32	2.27	2.22	2.18	2.15	2.10	2.05	1.99
27		**7.72**	**5.53**	**4.64**	**4.14**	**3.82**	**3.59**	**3.42**	**3.29**	**3.17**	**3.09**	**3.02**	**2.96**	**2.86**	**2.77**	**2.66**
		4.21	3.35	2.96	2.73	2.57	2.46	2.37	2.30	2.25	2.20	2.16	2.13	2.08	2.03	1.97
		7.68	**5.49**	**4.60**	**4.11**	**3.79**	**3.56**	**3.39**	**3.26**	**3.14**	**3.06**	**2.98**	**2.93**	**2.83**	**2.74**	**2.63**

Table 4 (*cont.*)

	Degrees of freedom: numerator														
Degrees freedom: denominator	1	2	3	4	5	6	7	8	9	10	11	12	14	16	20
28	4.20	3.34	2.95	2.71	2.56	2.44	2.36	2.29	2.24	2.19	2.15	2.12	2.06	2.02	1.96
	7.64	**5.45**	**4.57**	**4.07**	**3.76**	**3.53**	**3.36**	**3.23**	**3.11**	**3.03**	**2.95**	**2.90**	**2.80**	**2.71**	**2.60**
29	4.18	3.33	2.93	2.70	2.54	2.43	2.35	2.28	2.22	2.18	2.14	2.10	2.05	2.00	1.94
	7.60	**5.42**	**4.54**	**4.04**	**3.73**	**3.50**	**3.33**	**3.20**	**3.08**	**3.00**	**2.92**	**2.87**	**2.77**	**2.68**	**2.57**
30	4.17	3.32	2.92	2.69	2.53	2.42	2.34	2.27	2.21	2.16	2.12	2.09	2.04	1.99	1.93
	7.56	**5.39**	**4.51**	**4.02**	**3.70**	**3.47**	**3.30**	**3.17**	**3.06**	**2.98**	**2.90**	**2.84**	**2.74**	**2.66**	**2.55**
32	4.15	3.30	2.90	2.67	2.51	2.40	2.32	2.25	2.19	2.14	2.10	2.07	2.02	1.97	1.91
	7.50	**5.34**	**4.46**	**3.97**	**3.66**	**3.42**	**3.25**	**3.12**	**3.01**	**2.94**	**2.86**	**2.80**	**2.70**	**2.62**	**2.51**
34	4.13	3.28	2.88	2.65	2.49	2.38	2.30	2.23	2.17	2.12	2.08	2.05	2.00	1.95	1.89
	7.44	**5.29**	**4.42**	**3.93**	**3.61**	**3.38**	**3.21**	**3.08**	**2.97**	**2.89**	**2.82**	**2.76**	**2.66**	**2.58**	**2.47**
36	4.11	3.26	2.86	2.63	2.48	2.36	2.28	2.21	2.15	2.10	2.06	2.03	1.98	1.93	1.87
	7.39	**5.25**	**4.38**	**3.89**	**3.58**	**3.35**	**3.18**	**3.04**	**2.94**	**2.86**	**2.78**	**2.72**	**2.62**	**2.54**	**2.43**
38	4.10	3.25	2.85	2.62	2.46	2.35	2.26	2.19	2.14	2.09	2.05	2.02	1.96	1.92	1.85
	7.35	**5.21**	**4.34**	**3.86**	**3.54**	**3.32**	**3.15**	**3.02**	**2.91**	**2.82**	**2.75**	**2.69**	**2.59**	**2.51**	**2.40**
40	4.08	3.23	2.84	2.61	2.45	2.34	2.25	2.18	2.12	2.07	2.04	2.00	1.95	1.90	1.84
	7.31	**5.18**	**4.31**	**3.83**	**3.51**	**3.29**	**3.12**	**2.99**	**2.88**	**2.80**	**2.73**	**2.66**	**2.56**	**2.49**	**2.37**
42	4.07	3.22	2.83	2.59	2.44	2.32	2.24	2.17	2.11	2.06	2.02	1.99	1.94	1.89	1.82
	7.27	**5.15**	**4.29**	**3.80**	**3.49**	**3.26**	**3.10**	**2.96**	**2.86**	**2.77**	**2.70**	**2.64**	**2.54**	**2.46**	**2.35**
44	4.05	3.21	2.82	2.58	2.43	2.31	2.23	2.16	2.10	2.05	2.01	1.98	1.92	1.88	1.81
	7.24	**5.12**	**4.26**	**3.78**	**3.46**	**3.24**	**3.07**	**2.94**	**2.84**	**2.75**	**2.68**	**2.62**	**2.52**	**2.44**	**2.32**
46	4.05	3.20	2.81	2.57	2.42	2.30	2.22	2.14	2.09	2.04	2.00	1.97	1.91	1.87	1.80
	7.21	**5.10**	**4.24**	**3.76**	**3.44**	**3.22**	**3.05**	**2.92**	**2.82**	**2.73**	**2.66**	**2.60**	**2.50**	**2.42**	**2.30**
48	4.04	3.19	2.80	2.56	2.41	2.30	2.21	2.14	2.08	2.03	1.99	1.96	1.90	1.86	1.79
	7.19	**5.08**	**4.22**	**3.74**	**3.42**	**3.20**	**3.04**	**2.90**	**2.80**	**2.71**	**2.64**	**2.58**	**2.48**	**2.40**	**2.28**
50	4.03	3.18	2.79	2.56	2.40	2.29	2.20	2.13	2.07	2.02	1.98	1.95	1.90	1.85	1.78
	7.17	**5.06**	**4.20**	**3.72**	**3.41**	**3.18**	**3.02**	**2.88**	**2.78**	**2.70**	**2.62**	**2.56**	**2.46**	**2.39**	**2.26**
55	4.02	3.17	2.78	2.54	2.38	2.27	2.18	2.11	2.05	2.00	1.97	1.93	1.88	1.83	1.76
	7.12	**5.01**	**4.16**	**3.68**	**3.37**	**3.15**	**2.98**	**2.85**	**2.75**	**2.66**	**2.59**	**2.53**	**2.43**	**2.35**	**2.23**

Table 4 (cont.)

Degrees freedom: denominator	Degrees of freedom: numerator														
	1	2	3	4	5	6	7	8	9	10	11	12	14	16	20
60	4.00	3.15	2.76	2.52	2.37	2.25	2.17	2.10	2.04	1.99	1.95	1.92	1.86	1.81	1.75
	7.08	**4.98**	**4.13**	**3.65**	**3.34**	**3.12**	**2.95**	**2.82**	**2.72**	**2.63**	**2.56**	**2.50**	**2.40**	**2.32**	**2.20**
65	3.99	3.14	2.75	2.51	2.36	2.24	2.15	2.08	2.02	1.98	1.94	1.90	1.85	1.80	1.73
	7.04	**4.95**	**4.10**	**3.62**	**3.31**	**3.09**	**2.93**	**2.79**	**2.70**	**2.61**	**2.54**	**2.47**	**2.37**	**2.30**	**2.18**
70	3.98	3.13	2.74	2.50	2.35	2.23	2.14	2.07	2.01	1.97	1.93	1.89	1.84	1.79	1.72
	7.01	**4.92**	**4.08**	**3.60**	**3.29**	**3.07**	**2.91**	**2.77**	**2.67**	**2.59**	**2.51**	**2.45**	**2.35**	**2.28**	**2.15**
80	3.96	3.11	2.72	2.48	2.33	2.21	2.12	2.05	1.99	1.95	1.91	1.88	1.82	1.77	1.70
	6.96	**4.88**	**4.04**	**3.56**	**3.25**	**3.04**	**2.87**	**2.74**	**2.64**	**2.55**	**2.48**	**2.41**	**2.32**	**2.24**	**2.11**
100	3.94	3.09	2.70	2.46	2.30	2.19	2.10	2.03	1.97	1.92	1.88	1.85	1.79	1.75	1.68
	6.90	**4.82**	**3.98**	**3.51**	**3.20**	**2.99**	**2.82**	**2.69**	**2.59**	**2.51**	**2.43**	**2.36**	**2.26**	**2.19**	**2.06**
125	3.92	3.07	2.68	2.44	2.29	2.17	2.08	2.01	1.95	1.90	1.86	1.83	1.77	1.72	1.65
	6.84	**4.78**	**3.94**	**3.47**	**3.17**	**2.95**	**2.79**	**2.65**	**2.56**	**2.47**	**2.40**	**2.33**	**2.23**	**2.15**	**2.03**
150	3.91	3.06	2.67	2.43	2.27	2.16	2.07	2.00	1.94	1.89	1.85	1.82	1.76	1.71	1.64
	6.81	**4.75**	**3.91**	**3.44**	**3.14**	**2.92**	**2.76**	**2.62**	**2.53**	**2.44**	**2.37**	**2.30**	**2.20**	**2.12**	**2.00**
200	3.89	3.04	2.65	2.41	2.26	2.14	2.05	1.98	1.92	1.87	1.83	1.80	1.75	1.69	1.62
	6.76	**4.71**	**3.88**	**3.41**	**3.11**	**2.90**	**2.73**	**2.60**	**2.50**	**2.41**	**2.34**	**2.28**	**2.17**	**2.09**	**1.97**
400	3.86	3.02	2.62	2.39	2.23	2.12	2.03	1.96	1.90	1.85	1.81	1.78	1.72	1.67	1.60
	6.70	**4.66**	**3.83**	**3.36**	**3.06**	**2.85**	**2.69**	**2.55**	**2.46**	**2.37**	**2.29**	**2.23**	**2.12**	**2.04**	**1.92**
1000	3.85	3.00	2.61	2.38	2.22	2.10	2.02	1.95	1.89	1.84	1.80	1.76	1.70	1.65	1.58
	6.66	**4.62**	**3.80**	**3.34**	**3.04**	**2.82**	**2.66**	**2.53**	**2.43**	**2.34**	**2.26**	**2.20**	**2.09**	**2.01**	**1.89**
∞	3.84	2.99	2.60	2.37	2.21	2.09	2.01	1.94	1.88	1.83	1.79	1.75	1.69	1.64	1.57
	6.64	**4.60**	**3.78**	**3.32**	**3.02**	**2.80**	**2.64**	**2.51**	**2.41**	**2.32**	**2.24**	**2.18**	**2.07**	**1.99**	**1.87**

Source: Reprinted with permission from Iowa State University Press. The table appeared as table A14 part I on pages 476–479 of *Statistical methods*, eighth edition, by G. W. Snedecor and W. G. Cochran, 1989.

Table 5 Critical values of U, the Mann–Whitney statistic. Critical values of U and U' for a one-tailed test at $\alpha=0.005$ or a two-tailed test at $\alpha=0.01$

N_2 \ N_1	1	2	3	4	5	6	7	8	9	10	11	12	13	14	15	16	17	18	19	20
1	–	–	–	–	–	–	–	–	–	–	–	–	–	–	–	–	–	–	–	–
2	–	–	–	–	–	–	–	–	–	–	–	–	–	–	–	–	–	–	0	0
																			38	40
3	–	–	–	–	–	–	–	–	0	0	0	1	1	1	2	2	2	2	3	3
									27	30	33	35	38	41	43	46	49	52	54	57
4	–	–	–	–	–	0	0	1	1	2	2	3	3	4	5	5	6	6	7	8
						24	28	31	35	38	42	45	49	52	55	59	62	66	69	72
5	–	–	–	–	0	1	1	2	3	4	5	6	7	7	8	9	10	11	12	13
					25	29	34	38	42	46	50	54	58	63	67	71	75	79	83	87
6	–	–	–	0	1	2	3	4	5	6	7	9	10	11	12	13	15	16	17	18
				24	29	34	39	44	49	54	59	63	68	73	78	83	87	92	97	102
7	–	–	–	0	1	3	4	6	7	9	10	12	13	15	16	18	19	21	22	24
				28	34	39	45	50	56	61	67	72	78	83	89	94	100	105	111	116
8	–	–	–	1	2	4	5	7	9	11	13	15	17	18	20	22	24	26	28	30
				31	38	44	50	57	63	69	75	81	87	94	100	106	112	118	124	130
9	–	–	0	1	3	5	7	9	11	13	16	18	20	22	24	27	29	31	33	36
			27	35	42	49	56	63	70	77	83	90	97	104	111	117	124	131	138	144
10	–	–	0	2	4	6	9	11	13	16	18	21	24	26	29	31	34	37	39	42
			30	38	46	54	61	69	77	84	92	99	106	114	121	129	136	143	151	158
11	–	–	0	2	5	7	10	13	16	18	21	24	27	30	33	36	39	42	45	48
			33	42	50	59	67	75	83	92	100	108	116	124	132	140	148	156	164	172
12	–	–	1	3	6	9	12	15	18	21	24	27	31	34	37	41	44	47	51	54
			35	45	54	63	72	81	90	99	108	117	125	134	143	151	160	169	177	186
13	–	–	1	3	7	10	13	17	20	24	27	31	34	38	42	45	49	53	56	60
			38	49	58	68	78	87	97	106	116	125	125	144	153	163	172	181	191	200
14	–	–	1	4	7	11	15	18	22	26	30	34	38	42	46	50	54	58	63	67
			41	52	63	73	83	94	104	114	124	134	144	154	164	174	184	194	203	213
15	–	–	2	5	8	12	16	20	24	29	33	37	42	46	51	55	60	64	69	73
			43	55	67	78	89	100	111	121	132	143	153	164	174	185	195	206	216	227
16	–	–	2	5	9	13	18	22	27	31	36	41	45	50	55	60	65	70	74	79
			46	59	71	83	94	106	117	129	140	151	163	174	185	196	207	218	230	241
17	–	–	2	6	10	15	19	24	29	34	39	44	49	54	60	65	70	75	81	86
			49	62	75	87	100	112	124	148	148	160	172	184	195	207	219	231	242	254
18	–	–	2	6	11	16	21	26	31	37	42	47	53	58	64	70	75	81	87	92
			52	66	79	92	105	118	131	143	156	169	181	194	206	218	231	243	255	268
19	–	0	3	7	12	17	22	28	33	39	45	51	56	63	69	74	81	87	93	99
		38	54	69	83	97	111	124	138	151	164	177	191	203	216	230	242	255	268	281
20	–	0	3	8	13	18	24	30	36	42	48	54	60	67	73	79	86	92	99	105
		40	57	72	87	102	116	130	144	158	172	186	200	213	227	241	254	268	281	295

Note:
To be significant for any given N_1 and N_2, obtained U must be equal to or *less than* the value shown in the table. Obtained U' must be equal to or *greater than* the value shown in the table. *Example:* If $\alpha=0.01$, two-tailed test, $N_1=13$, $N_2=15$, and obtained $U=150$, we cannot reject $H0$ since obtained U is within the upper (153) and lower (42) critical values.
(Dashes in the body of the table indicate that no decision is possible at the stated level of significance.)

Table 5 (*cont.*) Critical values of *U* and *U'* for a one-tailed test at $\alpha=0.01$ or a two-tailed test at $\alpha=0.02$

N_2 \\ N_1	1	2	3	4	5	6	7	8	9	10	11	12	13	14	15	16	17	18	19	20
1	–	–	–	–	–	–	–	–	–	–	–	–	–	–	–	–	–	–	–	–
2	–	–	–	–	–	–	–	–	–	–	–	–	–	0	0	0	0	0	0	1	1
														26	28	30	32	34	36	37	39
3	–	–	–	–	–	–	–	0	0	1	1	1	2	2	2	3	3	4	4	4	5
								21	24	26	29	32	34	37	40	42	45	47	50	52	55
4	–	–	–	–	0	1	1	2	3	3	4	5	5	6	7	7	8	9	9	10
					20	23	27	30	33	37	40	43	47	50	53	57	60	63	67	70
5	–	–	–	0	1	2	3	4	5	6	7	8	9	10	11	12	13	14	15	16
				20	24	28	32	36	40	44	48	52	56	60	64	68	72	76	80	84
6	–	–	–	1	2	3	4	6	7	8	9	11	12	13	15	16	18	19	20	22
				23	28	33	38	42	47	52	57	61	66	71	75	80	84	89	94	93
7	–	–	0	1	3	4	6	7	9	11	12	14	16	17	19	21	23	24	26	28
			21	27	32	38	43	49	54	59	65	70	75	81	86	91	96	102	107	112
8	–	–	0	2	4	6	7	9	11	13	15	17	20	22	24	26	28	30	32	34
			24	30	36	42	49	55	61	67	73	79	84	90	96	102	108	114	120	126
9	–	–	1	3	5	7	9	11	14	16	18	21	23	26	28	31	33	36	38	40
			26	33	40	47	54	61	67	74	81	87	94	100	107	113	120	126	133	140
10	–	–	1	3	6	8	11	13	16	19	22	24	27	30	33	36	38	41	44	47
			29	37	44	52	59	67	74	81	88	96	103	110	117	124	132	139	146	153
11	–	–	1	4	7	9	12	15	18	22	25	28	31	34	37	41	44	47	50	53
			32	40	48	57	65	73	81	88	96	104	112	120	128	135	143	151	159	167
12	–	–	2	4	8	11	14	17	21	24	28	31	35	38	42	46	49	53	56	60
			34	43	52	61	70	79	87	96	104	113	121	130	138	146	155	163	172	180
13	–	0	2	5	9	12	16	20	23	27	31	35	39	43	47	51	55	59	63	67
		26	37	47	56	66	75	84	94	103	112	121	130	139	148	157	166	175	184	193
14	–	0	2	6	10	13	17	22	26	30	34	38	43	47	51	56	60	65	69	73
		28	40	50	60	71	81	90	100	110	120	130	139	149	159	168	178	187	197	207
15	–	0	3	7	11	15	19	24	28	33	37	42	47	51	56	61	66	70	75	80
		30	42	53	64	75	86	96	107	117	128	138	148	159	169	179	189	200	210	220
16	–	0	3	7	12	16	21	26	31	36	41	46	51	56	61	66	71	76	82	87
		32	45	57	68	80	91	102	113	124	135	146	157	168	179	190	201	212	222	233
17	–	0	4	8	13	18	23	28	33	38	44	49	55	60	66	71	77	82	88	93
		34	47	60	72	84	96	108	120	132	143	155	166	178	189	201	212	224	234	247
18	–	0	4	9	14	19	24	30	36	41	47	53	59	65	70	76	82	88	94	100
		36	50	63	76	89	102	114	126	139	151	163	175	187	200	212	224	236	248	260
19	–	1	4	9	15	20	26	32	38	44	50	56	63	69	75	82	88	94	101	107
		37	53	67	80	94	107	120	133	146	159	172	184	197	210	222	235	248	260	273
20	–	1	5	10	16	22	28	34	40	47	53	60	67	73	80	87	93	100	107	114
		39	55	70	84	98	112	126	140	153	167	180	193	207	220	233	247	260	273	286

Note:

To be significant for any given N_1 and N_2, obtained *U* must be equal to or *less than* the value shown in the table. Obtained *U'* must be equal to or *greater than* the value shown in the table. (Dashes in the body of the table indicate that no decision is possible at the stated level of significance.)

Table 5 (cont.) Critical values of U and U' for a one-tailed test at $\alpha=0.025$ or a two-tailed test at $\alpha=0.05$

N_2 \ N_1	1	2	3	4	5	6	7	8	9	10	11	12	13	14	15	16	17	18	19	20
1	–	–	–	–	–	–	–	–	–	–	–	–	–	–	–	–	–	–	–	–
2	–	–	–	–	–	–	–	0	0	0	0	1	1	1	1	1	2	2	2	2
								16	**18**	**20**	**22**	**23**	**25**	**27**	**29**	**31**	**32**	**34**	**36**	**38**
3	–	–	–	–	0	1	1	2	2	3	3	4	4	5	5	6	6	7	7	8
					15	**17**	**20**	**22**	**25**	**27**	**30**	**32**	**35**	**37**	**40**	**42**	**45**	**47**	**50**	**52**
4	–	–	–	0	1	2	3	4	4	5	6	7	8	9	10	11	11	12	13	13
				16	**19**	**22**	**25**	**28**	**32**	**35**	**38**	**41**	**44**	**47**	**50**	**53**	**57**	**60**	**63**	**67**
5	–	–	0	1	2	3	5	6	7	8	9	11	12	13	14	15	17	18	19	20
			15	**19**	**23**	**27**	**30**	**34**	**38**	**42**	**46**	**49**	**53**	**57**	**61**	**65**	**68**	**72**	**76**	**80**
6	–	–	1	2	3	5	6	8	10	11	13	14	16	17	19	21	22	24	25	27
			17	**22**	**27**	**31**	**36**	**40**	**44**	**49**	**53**	**58**	**62**	**67**	**71**	**75**	**80**	**84**	**89**	**93**
7	–	–	1	3	5	6	8	10	12	14	16	18	20	22	24	26	28	30	32	34
			20	**25**	**30**	**36**	**41**	**46**	**51**	**56**	**61**	**66**	**71**	**76**	**81**	**86**	**91**	**96**	**101**	**106**
8	–	0	2	4	6	8	10	13	15	17	19	22	24	26	29	31	34	36	38	41
		16	**22**	**28**	**34**	**40**	**46**	**51**	**57**	**63**	**69**	**74**	**80**	**86**	**91**	**97**	**102**	**108**	**111**	**119**
9	–	0	2	4	7	10	12	15	17	20	23	26	28	31	34	37	39	42	45	48
		18	**25**	**32**	**38**	**44**	**51**	**57**	**64**	**70**	**76**	**82**	**89**	**95**	**101**	**107**	**114**	**120**	**126**	**132**
10	–	0	3	5	8	11	14	17	20	23	26	29	33	36	39	42	45	48	52	55
		20	**27**	**35**	**42**	**49**	**56**	**63**	**70**	**77**	**84**	**91**	**97**	**104**	**111**	**118**	**125**	**132**	**138**	**145**
11	–	0	3	6	9	13	16	19	23	26	30	33	37	40	44	47	51	55	58	62
		22	**30**	**38**	**46**	**53**	**61**	**69**	**76**	**84**	**91**	**99**	**106**	**114**	**121**	**129**	**136**	**143**	**151**	**158**
12	–	1	4	7	11	14	18	22	26	29	33	37	41	45	49	53	57	61	65	69
		23	**32**	**41**	**49**	**58**	**66**	**74**	**82**	**91**	**99**	**107**	**115**	**123**	**131**	**139**	**147**	**155**	**163**	**171**
13	–	1	4	8	12	16	20	24	28	33	37	41	45	50	54	59	63	67	72	76
		25	**35**	**44**	**53**	**62**	**71**	**80**	**89**	**97**	**106**	**115**	**124**	**132**	**141**	**149**	**158**	**167**	**175**	**184**
14	–	1	5	9	13	17	22	26	31	36	40	45	50	55	59	64	67	74	78	83
		27	**37**	**47**	**57**	**67**	**76**	**86**	**95**	**104**	**114**	**123**	**132**	**141**	**151**	**160**	**171**	**178**	**188**	**197**
15	–	1	5	10	14	19	24	29	34	39	44	49	54	59	64	70	75	80	85	90
		29	**40**	**50**	**61**	**71**	**81**	**91**	**101**	**111**	**121**	**131**	**141**	**151**	**161**	**170**	**180**	**190**	**200**	**210**
16	–	1	6	11	15	21	26	31	37	42	47	53	59	64	70	75	81	86	92	98
		31	**42**	**53**	**65**	**75**	**86**	**97**	**107**	**118**	**129**	**139**	**149**	**160**	**170**	**181**	**191**	**202**	**212**	**222**
17	–	2	6	11	17	22	28	34	39	45	51	57	63	67	75	81	87	93	99	105
		32	**45**	**57**	**68**	**80**	**91**	**102**	**114**	**125**	**136**	**147**	**158**	**171**	**180**	**191**	**202**	**213**	**224**	**235**
18	–	2	7	12	18	24	30	36	42	48	55	61	67	74	80	86	93	99	106	112
		34	**47**	**60**	**72**	**84**	**96**	**108**	**120**	**132**	**143**	**155**	**167**	**178**	**190**	**202**	**213**	**225**	**236**	**248**
19	–	2	7	13	19	25	32	38	45	52	58	65	72	78	85	92	99	106	113	119
		36	**50**	**63**	**76**	**89**	**101**	**114**	**126**	**138**	**151**	**163**	**175**	**188**	**200**	**212**	**224**	**236**	**248**	**261**
20	–	2	8	13	20	27	34	41	48	55	62	69	76	83	90	98	105	112	119	127
		38	**52**	**67**	**80**	**93**	**106**	**119**	**132**	**145**	**158**	**171**	**184**	**197**	**210**	**222**	**235**	**248**	**261**	**273**

Note:

To be significant for any given N_1 and N_2, obtained U must be equal to or *less than* the value shown in the table. Obtained U' must be equal to or *greater than* the value shown in the table. (Dashes in the body of the table indicate that no decision is possible at the stated level of significance.)

Table 5 (*cont.*) **Critical values of U and U' for a one-tailed test at $\alpha=0.05$ or a two-tailed test at $\alpha=0.10$**

N_2 \ N_1	1	2	3	4	5	6	7	8	9	10	11	12	13	14	15	16	17	18	19	20
1	–	–	–	–	–	–	–	–	–	–	–	–	–	–	–	–	–	–	0	0
																			19	20
2	–	–	–	–	0	0	0	1	1	1	1	2	2	2	3	3	3	4	4	4
					10	12	14	15	17	19	21	22	24	26	27	29	31	32	34	36
3	–	–	0	0	1	2	2	3	3	4	5	5	6	7	7	8	9	9	10	11
			9	12	14	16	19	21	24	26	28	31	33	35	38	40	42	45	47	49
4	–	–	0	1	2	3	4	5	6	7	8	9	10	11	12	14	15	16	17	18
			12	15	18	21	24	27	30	33	36	39	42	45	48	50	53	56	59	62
5	–	0	1	2	4	5	6	8	9	11	12	13	15	16	18	19	20	22	23	25
		10	14	18	21	25	29	32	36	39	43	47	50	54	57	61	65	68	72	75
6	–	0	2	3	5	7	8	10	12	14	16	17	19	21	23	25	26	28	30	32
		12	16	21	25	29	34	38	42	46	50	55	59	63	67	71	76	80	84	88
7	–	0	2	4	6	8	11	13	15	17	19	21	24	26	28	30	33	35	37	39
		14	19	24	29	34	38	43	48	53	58	63	67	72	77	82	86	91	96	101
8	–	1	3	5	8	10	13	15	18	20	23	26	28	31	33	36	39	41	44	47
		15	21	27	32	38	43	49	54	60	65	70	76	81	87	92	97	103	108	113
9	–	1	3	6	9	12	15	18	21	24	27	30	33	36	39	42	45	48	51	54
		17	24	30	36	42	48	54	60	66	72	78	84	90	96	102	108	114	120	126
10	–	1	4	7	11	14	17	20	24	27	31	34	37	41	44	48	51	55	58	62
		19	26	33	39	46	53	60	66	73	79	86	93	99	106	112	119	125	132	138
11	–	1	5	8	12	16	19	23	27	31	34	38	42	46	50	54	57	61	65	69
		21	28	36	43	50	58	65	72	79	87	94	101	108	115	122	130	137	144	151
12	–	2	5	9	13	17	21	26	30	34	38	42	47	51	55	60	64	68	72	77
		22	31	39	47	55	63	70	78	86	94	102	109	117	125	132	140	148	156	163
13	–	2	6	10	15	19	24	28	33	37	42	47	51	56	61	65	70	75	80	84
		24	33	42	50	59	67	76	84	93	101	109	118	126	134	143	151	159	167	176
14	–	2	7	11	16	21	26	31	36	41	46	51	56	61	66	71	77	82	87	92
		26	35	45	54	63	72	81	90	99	108	117	126	135	144	153	161	170	179	188
15	–	3	7	12	18	23	28	33	39	44	50	55	61	66	72	77	83	88	94	100
		27	38	48	57	67	77	87	96	106	115	125	134	144	153	163	172	182	191	200
16	–	3	8	14	19	25	30	36	42	48	54	60	65	71	77	83	89	95	101	107
		29	40	50	61	71	82	92	102	112	122	132	143	153	163	173	183	193	203	213
17	–	3	9	15	20	26	33	39	45	51	57	64	70	77	83	89	96	102	109	115
		31	42	53	65	76	86	97	108	119	130	140	151	161	172	183	193	204	214	225
18	–	4	9	16	22	28	35	41	48	55	61	68	75	82	88	95	102	109	116	123
		32	45	56	68	80	91	103	114	123	137	148	159	170	182	193	204	215	226	237
19	0	4	10	17	23	30	37	44	51	58	65	72	80	87	94	101	109	116	123	130
	19	34	47	59	72	84	96	108	120	132	144	156	167	179	191	203	214	226	238	250
20	0	4	11	18	25	32	39	47	54	62	69	77	84	92	100	107	115	123	130	138
	20	36	49	62	75	88	101	113	126	138	151	163	176	188	200	213	225	237	250	262

Note:

To be significant for any given N_1 and N_2, obtained U must be equal to or *less than* the value shown in the table. Obtained U' must be equal to or *greater than* the value shown in the table.

(Dashes in the body of the table indicate that no decision is possible at the stated level of significance.)

Source: Reproduced with permission of the McGraw-Hill Companies. The table appeared as table I in *Fundamentals of behavioral statistics*, seventh edition, by R. P. Runyon and A. Haber, McGraw–Hill, 1991.

Table 6 Critical values of the chi-square distribution

Degrees of freedom df	P=0.99	0.98	0.95	0.90	0.80	0.70	0.50	0.30	0.20	0.10	0.05	0.02	0.01
1	.000157	.000628	.00393	.0158	.0642	.148	.455	1.074	1.642	2.706	3.841	5.412	6.635
2	.0201	.0404	.103	.211	.446	.713	1.386	2.408	3.219	4.605	5.991	7.824	9.210
3	.115	.185	.352	.584	1.005	1.424	2.366	3.665	4.642	6.251	7.815	9.837	11.341
4	.297	.429	.711	1.064	1.649	2.195	3.357	4.878	5.989	7.779	9.488	11.668	13.277
5	.554	.752	1.145	1.610	2.343	3.000	4.351	6.064	7.289	9.236	11.070	13.388	15.086
6	.872	1.134	1.635	2.204	3.070	3.828	5.348	7.231	8.558	10.645	12.592	15.033	16.812
7	1.239	1.564	2.167	2.833	3.822	4.761	6.346	8.383	9.803	12.017	14.067	16.622	18.475
8	1.646	2.032	2.733	3.490	4.594	5.527	7.344	9.524	11.030	13.362	15.507	18.168	20.090
9	2.088	2.532	3.325	4.168	5.380	6.393	8.343	10.656	12.242	14.684	16.919	19.679	21.666
10	2.558	3.059	3.940	4.865	6.179	7.267	9.342	11.781	13.442	15.987	18.307	21.161	23.209
11	3.053	3.609	4.575	5.578	6.989	8.148	10.341	12.899	14.631	17.275	19.675	22.618	24.725
12	3.571	4.178	5.226	6.304	7.807	9.034	11.340	14.011	15.812	18.549	21.026	24.054	26.217
13	4.107	4.765	5.892	7.042	8.634	9.926	12.340	15.119	16.985	19.812	22.362	25.472	27.688
14	4.660	5.368	6.571	7.790	9.467	10.821	13.339	16.222	18.151	21.064	23.685	26.873	29.141
15	5.229	5.985	7.261	8.547	10.307	11.721	14.339	17.322	19.311	22.307	24.996	28.259	30.578
16	5.812	6.614	7.962	9.312	11.152	12.624	15.338	18.418	20.465	23.542	26.296	29.633	32.000
17	6.408	7.255	8.672	10.085	12.002	13.531	16.338	19.511	21.615	24.769	27.587	30.995	33.409
18	7.015	7.906	9.390	10.865	12.857	14.440	17.338	20.601	22.760	25.989	28.869	32.346	34.805
19	7.633	8.567	10.117	11.651	13.716	15.352	18.338	21.689	23.900	27.204	30.144	33.687	36.191
20	8.260	9.237	10.851	12.443	14.578	16.266	19.337	22.775	25.038	28.412	31.410	35.020	37.566

Note:

The first column (df) locates each χ^2 distribution., The other columns give the proportion of the area under the χ^2 distribution that is above the tabled value of χ^2. The χ^2 values under the column headings of 0.05 and 0.01 are the critical values of χ^2 for $\alpha = 0.05$ and 0.01. To be significant,

$\chi^2_{obit} \geq \chi^2_{crit}$

Table 6 (cont.)

Degrees of freedom df	P=0.99	0.98	0.95	0.90	0.80	0.70	0.50	0.30	0.20	0.10	0.05	0.02	0.01
21	8.897	9.915	11.591	13.240	15.445	17.182	20.337	23.858	26.171	29.615	32.671	36.343	38.932
22	9.542	10.600	12.338	14.041	16.314	18.101	21.337	24.939	27.301	30.813	33.924	37.659	40.289
23	10.196	11.293	13.091	14.878	17.187	19.021	22.337	26.018	28.429	32.007	35.172	38.968	41.638
24	10.856	11.992	13.848	15.659	18.062	19.943	23.337	27.096	29.553	33.196	36.415	40.270	42.980
25	11.524	12.697	14.611	16.473	18.940	20.867	24.337	28.172	30.675	34.382	37.652	41.566	44.314
26	12.198	13.409	15.379	17.292	19.820	21.792	25.336	29.246	31.795	35.563	38.885	42.856	45.642
27	12.879	14.125	16.151	18.114	20.703	22.719	26.336	30.319	32.912	36.741	40.113	44.140	46.963
28	13.565	14.847	16.928	18.939	21.588	23.647	27.336	31.391	34.027	37.916	41.337	45.419	48.278
29	14.256	15.574	17.708	19.768	22.475	24.577	28.336	32.461	35.139	39.087	42.557	46.693	49.588
30	14.953	16.306	18.493	20.599	23.364	23.508	29.336	33.530	36.250	40.256	43.773	47.962	50.892

Source: Reprinted by permission of Addison Wesley Longman Ltd. The table appeared as table IV (Distribution of χ^2), on page 47 of *Statistical tables for biological, agricultural and medical researcher*, sixth edition, by R. A. Fisher and F. Yates, 1974.

Table 7 Critical values of *T* for the Wilcoxon signed-ranks test

	Level of significance for one-tailed test					Level of significance for one-tailed test			
	.05	.025	.01	.005		.05	.025	.01	.005
	Level of significance for two-tailed test					Level of significance for two-tailed test			
N	.10	.05	.02	.01	*N*	.10	.05	.02	.01
5	0	–	–	–	28	130	116	101	91
6	2	0	–	–	29	140	126	110	100
7	3	2	0	–	30	151	137	120	109
8	5	3	1	0	31	163	147	130	118
9	8	5	3	1	32	175	159	140	128
10	10	8	5	3	33	187	170	151	138
11	13	10	7	5	34	200	182	162	148
12	17	13	9	7	35	213	195	173	159
13	21	17	12	9	36	227	208	185	171
14	25	21	15	12	37	241	221	198	182
15	30	25	19	15	38	256	235	211	194
16	35	29	23	19	39	271	249	224	207
17	41	34	27	23	40	286	264	238	220
18	47	40	32	27	41	302	279	252	233
19	53	46	37	32	42	319	294	266	247
20	60	52	43	37	43	336	310	281	261
21	67	58	49	42	44	353	327	296	276
22	75	65	55	48	45	371	343	312	291
23	83	73	62	54	46	389	361	328	307
24	91	81	69	61	47	407	378	345	322
25	100	89	76	68	48	426	396	362	339
26	110	98	84	75	49	446	415	379	355
27	119	107	92	83	50	466	434	397	373

Note:
The symbol *T* denotes the smaller sum of ranks associated with differences that are all of the same sign. For any given *N* (number of ranked differences), the obtained *T* is significant at a given level if it is equal to or *less than* the value shown in the table. All entries are for the *absolute value* of *T*.

Source: Reproduced with permission of the McGraw-Hill Companies. The table appeared as table J₄ in *Fundamentals of behavioral statistics*, seventh edition, by R. P. Runyon and A. Haber, McGraw-Hill, 1991.

Table 8 Critical values of the Pearson correlation coefficient r

Degrees of freedom, $v = n - 2$	Level of significance for a one-tailed test			
	.05	.025	.01	.005
	Level of significance for a two-tailed test			
	.10	.05	.02	.01
1	0.988	0.997	0.9995	0.9999
2	0.900	0.950	0.980	0.990
3	0.805	0.378	0.934	0.959
4	0.729	0.811	0.882	0.917
5	0.669	0.754	0.833	0.874
6	0.622	0.707	0.789	0.834
7	0.582	0.666	0.750	0.798
8	0.549	0.632	0.716	0.765
9	0.521	0.602	0.685	0.735
10	0.497	0.576	0.658	0.708
11	0.476	0.553	0.634	0.684
12	0.458	0.532	0.612	0.661
13	0.441	0.514	0.592	0.641
14	0.426	0.497	0.574	0.623
15	0.412	0.482	0.558	0.606
16	0.400	0.468	0.542	0.590
17	0.389	0.456	0.528	0.575
18	0.378	0.444	0.516	0.561
19	0.369	0.433	0.503	0.549
20	0.360	0.423	0.492	0.537
21	0.352	0.413	0.482	0.526
22	0.344	0.404	0.472	0.515
23	0.337	0.396	0.462	0.505
24	0.330	0.388	0.453	0.496
25	0.323	0.381	0.445	0.487
26	0.317	0.374	0.437	0.479
27	0.311	0.367	0.430	0.471
28	0.306	0.361	0.423	0.463
29	0.301	0.355	0.416	0.456
30	0.296	0.349	0.409	0.449
35	0.275	0.325	0.381	0.418
40	0.257	0.304	0.358	0.393
45	0.243	0.288	0.338	0.372
50	0.231	0.273	0.322	0.354
60	0.211	0.250	0.295	0.325
70	0.195	0.232	0.274	0.302
80	0.183	0.217	0.256	0.283
90	0.173	0.205	0.242	0.267

Note:
n is the number of pairs.

Table 8 (*cont.*)

Degrees of freedom, $v = n-2$	Level of significance for a one-tailed test			
	.05	.025	.01	.005
	Level of significance for a two-tailed test			
	.10	.05	.02	.01
100	0.164	0.195	0.230	0.254
120	0.150	0.178	0.210	0.232
150	0.134	0.159	0.189	0.208
200	0.116	0.138	0.164	0.181
300	0.095	0.113	0.134	0.148
400	0.082	0.098	0.116	0.128
500	0.073	0.088	0.104	0.115

Source: Reprinted by permission of Addison Wesley Longman Ltd. The table appeared as table VII (The correlation coefficient) on page 63 of *Statistical tables for biological, agricultural and medical researcher*, sixth edition, by R.A. Fisher and F. Yates, 1974.